THE GREENING
OF
THE CITIES

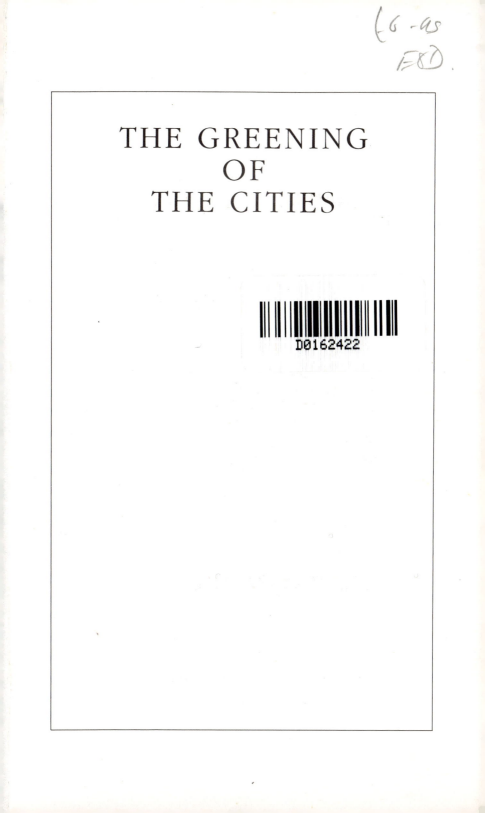

D0162422

THE GREENING
OF
THE CITIES

David Nicholson-Lord

ROUTLEDGE & KEGAN PAUL
LONDON & NEW YORK

First published in 1987 by
Routledge & Kegan Paul Ltd
11 New Fetter Lane, London EC4P 4EE

Published in the USA by
Routledge & Kegan Paul Inc.
in association with Methuen Inc.
29 West 35th Street, New York, NY 10001

Set in Garamond
by Columns of Reading
and printed in Great Britain
by T J Press (Padstow) Ltd
Padstow, Cornwall

Soc
HT
169
G7
N52
1987

Library of Congress Cataloging in Publication Data
Nicholson-Lord, David.
    The greening of the cities.
    (Geography, environment, and planning series)
    Bibliography: p.
    Includes index.
    1. City planning – Environmental aspects –
Great Britain.   2. Urban beautification – Environmental
aspects – Great Britain.   3. Human ecology – Great Britain.
4. Urban ecology (Biology) – Great Britain.   I. Title.
II. Series.
HT169.G7N52   1987   307.1'2   87–12800

British Library CIP Data also available
ISBN 0–7102–0328–4 (p)

For Jessie, Sue and Katy

CONTENTS

FIGURES

PLATES

PLATE SECTION II

FOREWORD

The majority of our large manufacturing cities are in decline – thousands of acres of their former industrial greatness have become gigantic scrapheaps. New industries with new technologies no longer make it necessary to locate industry in cities, and social and fiscal pressures are drawing people out into the countryside.

Thus a conflict is growing with cities dying for lack of industry and new housing – whilst conservationists resist the spread of development into the green belts or further into the rural landscape.

Yet perhaps this presents an opportunity to use waste land in cities to bring back the green space they have lacked since the industrial revolution whilst restoring to smaller towns and villages the vitality they lost with the drift of people to the cities to provide the labour force.

We urgently need a land-use policy which looks simultaneously at the towns and country and strikes a balance between urban and rural renewal.

I welcome this book as an important contribution to a contemporary debate which is of significance to everyone living in Britain who cares about its future.

Franklyn Perring
General Secretary

Royal Society for Nature Conservation
Lincoln
27 July 1987

PREFACE

Over the last two decades cities throughout much of the industrialized western world have undergone the beginnings of a great transformation, a transformation with its roots in demographic, technological and cultural processes. These processes have generally escaped the headlines although many of the symptoms – most notably repeated inner-city disturbances – have not. Coinciding with this transformation has been the rise of a movement known, broadly and baldly, as environmentalism, culminating in the 1980s in the emergence of the 'green' factor at the forefront of political debate, in Britain and elsewhere.

In a limited sense this book is about the meeting – in many instances 'collision' might be a more appropriate term – of these two forces: hence its title. But just as 'city' is an unsatisfactory shorthand for an elusive reality, so environmentalism is far more than a green party, far more even than an increasingly influential lobby. It is a philosophy, a mood, a way of viewing the world and thus of thinking, feeling and acting. As such, its dislocative and regenerative effects, although they may be more striking, more obviously 'revolutionary', in cities, will spill over into other patterns of settlement and land-use. The reverberations of the collision between environmentalism and the city can thus be heard in the countryside, not merely our own domestic countryside of green belt and national park, but the wider global countryside of the Third World where, as the final chapter of the book indicates, the odd and frequently tortured story of men's attitudes to civilisation and wilderness, to town and country, is being replayed on a grand, alarming and quite possibly fateful scale.

These paragraphs are, in a sense, a kind of apology – for the structure of a book that is about cities, landscaping, planning and so forth and yet strays into some unfamiliar terrain. In reality, no apology is necessary. What happens in the city and what happens in the world outside it are intimately linked, not only because cities and countryside are both, fundamentally, patterns of land-use but also because civilisations continue – just – to be most urgently made in cities. To rephrase a famous remark, what London does today, the rest of the world does tomorrow. In that sense, the greening of the cities is in the nature of a large-scale social experiment, where the very starkness of the issues involved has afforded a unique insight into the forces shaping the way we live and the way (and the places in which) we are going to go on living. This, I hope, explains the title of the last chapter, 'Beyond the city'.

Two other points deserve mention. The first is the vexed question of environmental determinism. Which comes first, poor people or bad places? Clearly no simple answer is possible – but equally clearly, as emerges constantly throughout the book, there is, and there has always been, a strong relationship and a vital interaction between people and place. As to whether the environmental psychologists and the human geographers will ever elaborate this relationship into a convincing theorem, only time – probably a very long time – will tell.

The second point concerns the treatment of London, particularly in the historical sections, as in some sense a representative city. Partly for the reasons advanced at the start of Chapter 1, I believe this is a reasonable approach. Few other cities combine such longevity and such complex yet typical patterns and cycles of development with such a dominant role in pre-industrial, industrial and (prospectively) post-industrial worlds.

ACKNOWLEDGMENTS

Much of the initial research for *The Greening of the Cities* was done during a three-month press fellowship at Wolfson College, Cambridge, in early 1983, under a programme launched the previous autumn. My thanks go, first, to *The Times* for allowing me leave of absence; to friends at Wolfson, in particular Bill Kirkman, for their welcome, company and ideas; and to British Petroleum which, with the Nuffield Foundation, sponsors the press fellowship programme and has also given generous support to the book. BP's role, coincidentally, is typical of the increased community involvement on the part of many British companies which is described in the book. I must congratulate them on their acuity: if ever an institution needed greater community involvement, and the openness to new ideas and influences that comes with it, it is the British press. The programme richly deserves to succeed and I hope this book may do something to help.

Secondly, I should like to thank members of several Cambridge University faculties, particularly geography and land economy, for sparing many hours to talk to me, and staff at the Landscape Institute for help with research. I am especially grateful to Michael Chisholm, Bob Bennett and Derek Nicholls at Cambridge and to Sheila Harvey at the Institute. Heartfelt thanks also to Sue for valiant and selfless efforts with a truly appalling manuscript and to Katy J. for reminding me that there are things other than work.

Four organizations have lent their support to this book: the Royal Society for Nature Conservation – representing the county naturalists' trusts – whose general secretary, Dr Franklyn Perring, has written the foreword; the National Council for Voluntary Organizations; the Civic Trust; and the Think Green campaign. Their backing is much appreciated and highly appropriate, since all four bodies embody the spirit of environmental voluntarism – the commitment to one's own neighbourhood and the straightforward desire to make it look and feel better – which is at the heart of the greening of the cities. Yet although each has played an important role in the greening movement, the movement has fundamentally been one of people rather than organizations. Finally, therefore, I should like to express my respect and admiration for all those thousands of people who, against many odds and often with little recognition, are helping to remake our cities. At the risk of sounding trite, they are the real heroes of this book, not least because they have done things rather than thought about doing things or talked about doing things or waited for somebody else to do things. As a city-dweller through birth, necessity and a peculiar kind of principle, I owe them – and so do many millions more who perhaps do not realize it – a great debt of gratitude.

CREDITS

I am also grateful to the following: to Faber and Faber and Harper & Row for permission to quote the extract from 'Pike', from *Lupercal*, by Ted Hughes. To Faber and Faber and Farrar, Straus and Giroux for permission to quote the extract from 'Going, Going', from *High Windows*, by Philip Larkin.

Acknowledgments are due to the following for photographs: between pages 56–57: 1, author; 2, Landlife; 3, Robert Tregay; 4, British Trust for Conservation Volunteers; 5, author; 6.1, Urban Wildlife Group; 7.1, 7.2, author; 8.1, 8.2, Landlife; 8.3, Robert Tregay; 8.4, 8.5, author; 9.1, British Trust for Conservation Volunteers (BTCV); 9.2, BTCV, Don Williams, Shell UK Ltd; 9.3, Joan McCarthy, Benwell Nature Park; 10.1, 10.2, Lin Whitfield, National Federation of City Farms; between pages 162–163: 1, author; 2, Sun Alliance/Land-use Consultants (Ian the Ruthven); 3, Leicester City Wildlife Project; 4, Tyne and Wear council; 5, Scottish Development Agency, Scottish Special Housing Association; 6, Robert Tregay 11.1, 11.2, Greater Manchester Council; 11.3, author; 11.4, City of Bristol; 11.5, Scottish Development Agency, Scottish Special Housing Association; 13.1, Milton Keynes Development Corporation; 13.2, Robert Tregay.

For Figure 13.1, acknowledgments are due to Derek Walker Associates.

1

DETRITUS

Its urbanization, progressing steadily, had finally reached the ultimate. All the land surface of Trantor, 75,000,000 square miles in extent, was a single city . . .. It had only one function, administration; one purpose, government; and one manufactured product, law . . .. There was no living object on its surface but man, his pets and his parasites.

(The imperial planet-city of Trantor, described in Isaac Asimov's 'Foundation Trilogy', 1951–3)

At some point in the early 1960s an event took place the significance of which was not fully appreciated until well over a decade later. London – that is, the Greater London area – began to lose jobs. Its rate of growth had been slowing, in common with other big cities, since the 1950s but so far the loss of employment had been relative: it had merely grown more slowly. In 1961 or 1962, the loss became absolute.

Greater London is, of course, a useful administrative fiction. London's true identity has been in doubt ever since the king and court set up a rival enterprise at Westminster in the eleventh century. Nevertheless it is legitimate to describe the event as a watershed. London led the west into a new era of urban growth and for almost a century, until overtaken by New York, was the largest city in history. It was the first of the 'cities of the main street of the world', in Robert Park's vivid phrase, and continues to embody so much of what we mean by the word 'city'. Indeed throughout much of the reign of the Tudor, Stuart and Hanoverian monarchs, as Raymond Williams has pointed out, London was *the* city. The rest of England, even thriving provincial centres like Norwich and Bristol, was merely 'country'.

London's employment had grown, with its population, for the best part of a millennium. The abrupt reversal of its growth, and that of the other big cities, caught Government on the hop and lends a curiously old-fashioned air to many works on urbanization written before 1970. These acknowledge some decentralization but assume, for most of us, a relatively closely packed urban future, hence the flutterings at the census results of 1971 and 1981, which showed an overall drop in London's population of 16 per cent since 1961. More than a million and a quarter people moved out of the capital in those two decades, leaving London's population below seven million for the first time since 1901. In the 1970s, moreover, the rate of outflow was accelerating. Over the same period, meanwhile, another three-quarters of a million people, or 17.5 per cent of 1961 populations, moved out of the six principal English cities – Birmingham, Leeds, Liverpool, Manchester, Newcastle-upon-Tyne and Sheffield. The inner cities were the worst hit. Inner London lost one million people – almost a third of its population – between 1961 and 1981.

The effect on jobs was devastating. Since the mid-1960s the employment base of the big cities has collapsed, initially and most acutely in manufacturing but later extending to services and office-based jobs. Between 1960 and 1981, London and the major conurbations lost 1.7 million manufacturing jobs, 79 per cent of the total national loss of 2.1 million jobs. By 1985 London could claim the dubious distinction of having the largest concentration of unemployed people in the advanced industrial world. In relative terms, however, the great northern cities like Manchester, Liverpool and Newcastle were much worse off.

Government, notably the Labour administration of 1974–79, responded by dusting off a somewhat antiquated urban programme – founded, remarkable though it now seems, as long ago as 1968 – and declaring an end to the long-running New Towns programme accused (wrongly) of siphoning off inner-city investment. It also abolished the Location of Offices Bureau, although offices continued, at a somewhat diminished rate, to leave London and the other cities. The inner cities, meanwhile, denuded progressively of jobs and people and erupting into riots in 1981 and 1985, have assumed a status in contemporary myth rather more alarming than the depressed areas of the 1930s.

This brief summary only hints at the more significant implications of the emptying of the cities. The first is that it predates the major recession beginning in the mid-1970s and is generally acknowledged to have its deepest roots elsewhere. Indeed it seems likely that city decline has become a product of economic growth: an upturn in the economy would thus intensify it.

Linked with this is a second important feature, which is that the extent of decline is directly related to the extent of urbanization. The fastest growth tends to be associated with the most 'rural' places. The bigger the city, meanwhile, the more quickly it is emptying. The representative modern growth point is no longer metropolis but, to use one recent American neologism, 'micropolis'.

A leading geographer, Brian Berry, has employed the rather more precise term 'counter-urbanization', a phenomenon first described in the United States in the mid-1970s – Berry's usage dates from 1976 – and then diagnosed further afield. In 1982 a study found evidence of it in nine out of fourteen European countries. In 1983 a manufacturing shift from cities to rural areas was documented for the first time throughout the European Economic Community. For whatever reason, counter-urbanization appears to affect developed countries *in proportion to their development*.

The rural renaissance has produced some surprising results. One of America's most inhospitable tracts of sparseland, the Great Plains plateau east of the Rocky Mountains, has seen a reversal of its long-standing population drain. And in Britain growth has been fastest in regions like East Anglia and the South-West, where the population increase over the two decades 1961–81 was, respectively, 27 and 18 per cent. Cornwall's population, for example, grew by a quarter between 1961 and 1981. A survey of these immigrants to Cornwall found them to be economically active – typically aged between 40 and 60 – and generally very well educated. Many also seemed to be unemployed, at least formally, although they may instead have been devoting themselves to that rich mixture of do-it-yourself, barter, exchange and payment in kind known as the 'black economy'. They appeared, the survey concluded, to be searching for an alternative lifestyle.

This sketchy portrait of the new urban refugees – 'drop-outs' is too idiosyncratic a term in view of the numbers involved – touches on two other features of the exodus. The first is its relationship with the flight to the suburbs: there are links but there are also important differences. The second is that, having served for so long as a brake on the movement of people out of cities or as an inertial force preventing them from moving too far, industry has now enthusiastically joined in the flight. Indeed it is well in its van.

Although the picture thus emerges of jobs now leaving the cities faster than people, the phenomenon is, of course, less an actual physical flight than the death of old firms in the city and the birth of new ones elsewhere. More pliant rural workforces, spiralling urban rents and rates, the growth of 'footloose' industry able to operate from rural seclusion by means of information technology – all these have been suggested as causes. So, too, has the concentration of ageing industry in the cities. Yet there is clearly more involved. The EEC-wide analysis already mentioned points to urban regions broadly enjoying a more modern spread of industry than rural – a structural pattern which should have produced a shift from country to city, not the reverse. Forces are indicated, it seems, 'whose strength is directly dependent on the scale and degree of urban development, measured by size and density of urban agglomeration'. A strong candidate may be a lack of land in cities on which space-hungry businesses, increasingly demanding more square feet per worker, can expand. This is the influential 'constrained location' theory advanced by a Cambridge University research team.

Yet the freeing of industry from its need to be near the city and what it offered – transport, services, concentrations of labour – has also enabled it to respond far more to the values of its workforce. Suburbanization, the modern quest for arcadia, offers one historic clue to those values but there are signs that this concept itself is undergoing a fundamental change – that the contemporary vision of arcadia is altogether earthier, more muscular, wilder and in some respects more resolutely anti-urban than its precursors. It is this vision of arcadia, and the changes it is enforcing on cities, which forms much of the matter of this book.

The cities are meanwhile suffering the backwash of the rural revival. The exodus has coincided with the progressive collapse of the urban foundations built by the Victorians: the sewers, drains and watermains, and also the houses. In 1985 the country faced a housing repairs bill estimated at £47 billion. Technological and industrial change has meant that great lumps of the city's fabric have suddenly become so much scrap. 'Smokestack' industries are vanishing; containerization has superannuated the docks. North Sea gas has made the gasometer redundant, while electricity generation has shifted to isolated coastlines. Never before in recorded history has a civilization left such vast junkyards behind it as it passes on to found new settlements. It is, in a sense, prehistoric 'slash and burn' on a massive scale.

The consequence is the freeing of space in unprecedented quantities. The chief feature of dereliction in the 1970s was that, from being a phenomenon of mining areas and mineral workings, it arose suddenly and spectacularly in the cities. The geographer Alice Coleman, one of the few people to have examined in detail land-use changes in Britain since the Second World War, found in the Inner London borough of Tower Hamlets a threefold rise in what she calls 'dead and disturbed space' between 1964 and 1977. In the latter year it amounted to 14 per cent of the borough. Other statistics more starkly convey the sense of unproductive emptiness and neglect in

Tower Hamlets: the 30 miles of corrugated iron, the amount of land – 37 per cent of the total – where buildings were too sparse for the term 'townscape' to be applied, the fact that 56 per cent of the borough generated no rate income.

The old industrial towns and cities have suffered the worst. Vacant land in the eastern area of Glasgow, for instance, has been estimated at 20 per cent. In Greater Manchester alone, in 1982 the officially registered area of derelict land was five times greater than in the whole of East Anglia and 60 per cent more extensive than in all twelve rural counties in the South-East – an area twenty times as large. Manchester has been active in reclamation yet such has been the pace of decay that while 4,200 acres were restored between 1974 and 1982, another 5,760 fell derelict.

The true extent of urban waste space in Britain can only be guessed at. In 1982 Greater London and the six main conurbations, occupying 6.5 per cent of the land area of England, contained 32 per cent of the country's total dereliction. This total was put at 113,000 acres, an area substantially larger than the Isle of Wight. One of the few certainties about this figure, however, is that it is a gross underestimate. The local authorities responsible for the derelict land returns, for example, count differently from each other. They also ignore, at Government behest, land which most people would regard as derelict – small sites awaiting development, for instance. Much urban dereliction comes in tiny pockets of land and results from comprehensive redevelopment schemes which never materialize. One survey of Birmingham found that two-thirds of the empty land in its centre consisted of sites of less than five acres. Thirty-five years after the war ended, a bomb-site in Hull was still empty: in 1947 the council had refused residents a community hall there because the land was needed for redevelopment. Such examples could be repeated interminably. Hence it is no surprise that a council survey in the West Midlands in 1974 counted more than three times as much dereliction as recorded in official figures.

The tininess of so many sites, testimony to an erosion of the city almost geological in its gradualness, goes some way to explaining the peculiar invisibility of the issues at stake, particularly in Westminster, the City and Fleet Street. It takes imagination, fact-finding tours or a riot to remove the mental cosseting of the metropolitan commuter.

Even less visible, however, is the vacancy behind office façades and factory windows. Between 1979 and 1983 the national stock of available industrial floorspace more than tripled. Available office space in Britain almost doubled between 1981 and 1984. Such hidden emptiness is the dereliction of tomorrow. A third of the 153 million square feet of vacant industrial floorspace in England and Wales in 1984 was estimated to be 'chronically unlettable'. Office blocks erected in haste only a decade earlier stood as monuments to a vanished seller's market, when city-centre property was a blue-chip investment. Inflexibly designed, expensive to run, lacking the space for cabling required by new technology, they were the commercial equivalent of high-rise council estates. In 1982 a watershed was reached when the total area of industrial floorspace began its decline from the historical peak of the previous year. At this stage, vacant industrial and office space together amounted to at least 200 million square feet, the equivalent of 1,000 Centrepoints. The change from concealed to revealed emptiness, however – from old buildings to new uses (or no uses at all) – was now showing through in the statistics.

Houses must be included in this picture of 'hidden' emptiness. Between 1977 and 1985 the number of vacant houses in London and the six metropolitan counties alone grew by 24 per cent to reach a figure of 294,000, equivalent to over 23,000 acres of unused land – an area the size of the city of Coventry. Since seven-tenths of the houses were privately owned, many had clearly been abandoned. The 1981 census 'snapshot' produced a figure of 974,000 empty dwellings for the United Kingdom, equal to 96,000 acres of waste land, almost as much again as the official total of derelict land for the whole of England.

Emptiness in the cities is thus clearly much greater than official figures imply. The Civic Trust in 1977 put the extent of urban wasteland at 250,000 acres. In *Britain's Wasting Acres*, the land-use planner Graham Moss estimated the national total of spoiled, blighted and idle land to be 2.5 million acres, an area the size of Devon and Cornwall. Since there is no national land-use survey, one has to make do with estimates like these, however unsatisfactory.

The failure to put this land to new use is a critical factor in the decline of the cities. Some of the reasons are physical: the sheer volume of clutter and debris, the poisoning of the soil by industrial pollutants, the costs of large-scale reclamation and 'detoxification'. Much initiative has been thwarted by the immense complexities of land ownership: it has often proved impossible to find out who actually owns a particular patch of ground. One expert working party concluded that with much disused land, developers would make a loss even if they had acquired it for nothing, so corrosive has been a century and more of industrial habitation.

Many landholders also refuse to entertain the possibility – publicly, at least – of a drop in land values. Overall it seems that well over half the waste land is publicly owned, 60 per cent of this by councils and about a quarter by nationalized or state industries. The urban land market thus assumes a stagnant, monopolistic aspect. In London's docklands the creation of a development corporation has sheared straight through the complexities and turgidities of the land market. Development pressures throughout the South-East have also, though somewhat patchily, spurred on land take-up in the capital. Elsewhere, however, the land market continues to act as a brake on the emergence of a new city.

Yet there is an important sense in which the ravelled intricacies of the land market stand as a metaphor for a deeper blockage. In *The Four-Gated City*, the final part of Doris Lessing's fictional account of the twentieth century, 'Children of Violence', the central character, newly arrived in post-war Britain from Africa, watches excavations in London. She

> had never before seen soil that was dead, that had no roots. How long had this street been built? … if one was to wade through earth in Africa, around one's legs roots: tree roots, thick, buried branches … a mat of working life … But walking here, it would be through unaired rootless soil, where electricity and telephone and gas tubes ran and knotted and twined.

The impression is partly of the city, indeed of civilized life, as a lie, built on dearth and deadness. But there is also a powerful feeling of exhaustion through *over habitation*: a place has become too cluttered and enmeshed in its past. It recalls the strange observation of Strehlow on the aboriginal Arunta tribes of Australia, for whom every landscape feature has a mythical association. 'Their forefathers,' he remarked, 'have left them not a single unoccupied scene which they could fill with creatures of their own imagination.' Tradition had stifled creativity, no new myths were being invented. 'They are on the whole … not so much a primitive as a decadent race.' It also brings to mind Jane Jacob's argument in *The Economy of Cities* that initiative and innovation are products of disordered, not regimented, environments.

Regimentation and overelaborateness are hallmarks of institutions that have outlived their original function and are losing touch with reality. Amongst those caught up in their trails, they can induce an unreasoning desire for escape – physical flight, perhaps, or a bonfire of controls. The revival of the smaller country towns in Tudor and Stuart times, based in many areas on a spontaneous growth of local markets, was in part a response to the dead hand of the urban craft guilds and corporations: in their insistence on regulating every commercial activity, they merely succeeded in killing it off or driving it outside the city. Too many inner-city local authorities have displayed a similar perversity, imposing increasing restrictions on a diminishing amount of activity. Hence the urban development corporation, the enterprise zone and the free port – applications of an axe to the Gordian knot.

Yet no amount of purely administrative deregulation could abolish the boundary crossed by Doris Lessing's heroine, between African bush and London street, because they represent the separate worlds of country and city, one of which possesses the mysterious quality of 'life' while the other does not. And these feelings are extraordinarily pervasive, even amongst the supposedly dispassionate. A study of private housebuilders found that a major reason deterring them from even contemplating building on empty city land was the city's image: not only of a poor environment but of delays, restrictions and complexities. The difficulties were often entirely imaginary. Yet a green field site was, in some unspecified and ambiguous way, simpler, cleaner and freer.

Such feelings are rooted deep in the flight to the suburbs. When Peter Hall and colleagues surveyed householders on their 'desired future place' of residence in the early 1970s, there was an overwhelming presumption – not merely a wish – that movement would be outwards from the city. Fifty-nine per cent of the respondents listed 'country', 29 per cent suburbs and 8 per cent towns. Yet the fact that Hall's families were *actually* more likely to end up in a suburb rather than the country – the figures were 58 per cent and 35 per cent – indicates a considerable unrealized aspiration. For many the suburbs are clearly only a staging post in a longer journey.

How much is 'pull' and how much is 'push' is a moot point. Large-scale studies for the International Labour Office and the US Bureau of Census show that living conditions, crime, pollution and congestion worsen as cities grow larger and that dissatisfaction with neighbourhood is directly related to size. Inhabitants of cities of over three million are between four and seven times more likely to be dissatisfied than those of towns and rural areas of under 50,000 – a preference apparently now reflected in the link established between an area's speed of decay and the extent of its urbanization. A MORI poll in 1983 found the two most important factors listed as contributing to the quality of life were safe streets, cited by 72 per cent, and attractive countryside, mentioned by 53 per cent. Both, self-evidently, bear directly on what we now mean by city and country.

These are a few studies from the many available. What they speak of is the increasingly exclusive identification of the countryside with the qualities necessary for a fulfilling and civilized life: the city, inevitably, is left with the dregs. And this is also how the countryside sees and sells itself to the industrialists and entrepreneurs of tomorrow. Derbyshire is a 'beautiful place to grow', King's Lynn a place to 'live and work and breathe'. In Macclesfield you can 'be yourself', have enough room to develop your 'own kind of lifestyle'.

This is the language of industrial promotion. It is also the language of psychotherapy and its theme, ultimately, is freedom – the physical, and also the psychological, space to be as one wants. The irony is, of course, that for much of history it was the city which offered emancipation and escape from oppression. For the medieval serf fleeing his feudal bonds, the city was an oasis of liberty. That it should now be cast for so many in the role of oppressor is a sad and remarkable postscript, although one dare not yet call it an epitaph.

2

THE GREAT WEN VERSUS
THE GARDEN CITY

Cities ... have not always existed; they began at a certain period in the evolution of society and can equally be ended or radically transformed at another. They came into being not as a result of any natural necessity, but as the result of an historical need, and they will continue only for as long as this need persists.

(Leonardo Benevolo, The History of the City, *1980)*

Roman Londinium, in its high imperial summer the best part of two millennia ago, numbered at most 50,000 people and was a mile across. By William the Conqueror's time its population had sunk to 15,000. In 1777, when Dr Johnson remarked that the man who was tired of London was tired of life, open country was two miles from the centre of town and there were probably 750,000 inhabitants. Twentieth-century Greater London, at its height, held over eight and a half million people and covered 610 square miles. Its built-up area in parts was up to thirty miles across.

Which is the true London? Do all, equally, deserve to be called 'city'? Was a Roman Londoner as 'urban' as a Greater Londoner? If cities are in truth a result of historical need, at what point in their life-cycle does alteration become transformation? How shall we know when the city has ended?

Such questions go to the heart of Western society and how it defines itself. As will be argued in this chapter, they touch on issues fundamental to the development of the city, particularly in its phase of headlong industrial growth – issues of centrality and identity, and, by extension, of urban containment and rural conservation: the invention of the green belt has served as a particular focus. Slowly over the twentieth century two broadly divergent schools of thought have emerged: one, essentially conservative, treating the concentration of population in identifiable (and thus controllable) centres as politically convenient or historically inevitable; the other, rooted in what might be styled the unfolding 'folk culture' of ecology but taking many of its cues from technological change, feeling its way towards new, broadly decentralized living patterns. Yet the argument between them remains largely unfocused, in part because cities are large and complex forms and their changes are measured over decades, even centuries: people's memories are too short to cope, their vocabularies inadequate.

In 1890 William Morris solved this problem by projecting his city-watcher into the twenty-first century: he goes to sleep in London present to wake in London future and finds the city, and society as a whole, altered beyond belief. *News from Nowhere*, traditionally regarded as a Utopian romance, is coming to rank as one of the most inspired pieces of prediction produced by any of the great Victorians, Wells included. Its significance is twofold. Many of those now engaged in the grass-roots redesign of cities stand squarely in the tradition represented by Morris, sharing his hopes and his visions: there is thus an uncanny element of self-fulfilling prophecy about *News from Nowhere*. And in its imaginative telescoping of the process of change, it helps to isolate and clarify those elements which have to exist before a city can be said to exist.

The main cause of the transformation of the city in *News from Nowhere* was, we learn, the 'exodus of the people from the town to the country, and the gradual recovery by the town-bred people on one side, and the country-bred people on the other, of those arts of life which they had each lost'. In London the Thames once again has salmon; Trafalgar Square is an apricot orchard and rose gardens bloom off Shaftesbury Avenue. London, indeed much of the country, has returned to woodland and forest, pieces of 'wild nature' are maintained because they are both enjoyed and needed and a 'passionate love of the Earth', somewhat un-Victorian, has become widespread. Parliament is a dung market.

London, however, assumes an indeterminate identity. There are 'noble' buildings, with hints of Renaissance Florence and Byzantium, but they remain just that – collections of buildings. There are elegant shopping arcades – Piccadilly survives as such. But the matrix of woodland, garden, field and orchard in which they are set means that they no longer cohere as a distinctive entity. The prevailing impression, instead, is of a distinguished and unusually fecund suburbia. And the lure of the city has faded. Told that country people are occasionally apt to cluster in Piccadilly, the narrator remarks, 'I couldn't help smiling to see how long a tradition would last. Here was the ghost of London still asserting itself as a centre – an intellectual centre, for aught I knew.'

It is in this casual comment about the ghost of London that Morris touches a vital cultural resonance – the preoccupation of civilized western society with the forms of urbanism, and its corresponding inability to rid itself of the idea of the city.

Cities have a shape but they also have a meaning, and from this deceptively simple duality much confusion has arisen. Take Braudel's definition of a town, for example – an 'unusual concentration of men and houses, close together, often joined wall to wall'. One does not have to cite the once-fashionable juristic definitions of cities as seats of episcopal sees or holders of royal charters to realize that Braudel's definition would be inadequate for a city because it does not render that extra symbolic dimension. These two halves of the city's character – its physical, spatial form and its cultural significance – conjoin in the city centre. Centrality, the burden of symbolism it carries and the extent to which that symbolism commands popular assent, are thus key determinants of the form and function of cities.

Throughout much of history the 'greatness' of cities has consisted in their expression of centralist ideologies – ideologies which tend towards the concentration of powers or qualities in one person or one place. Lewis Mumford, discussing the earliest civilizations of Egypt and Mesopotamia, describes founding cities as the 'special and all but universal function of kings'. And since kingship was 'lowered down from heaven', in the words of the Sumerian kinglist, cities became the home of the gods.

The same principle is evident at a more obviously political level. A phenomenon of London's late medieval growth was its distancing of every other British town. In the two centuries up to the reign of Henry VIII (1509–47) its wealth grew from three times to ten times that of its nearest competitor, and in the following two centuries its population increased from five to fifteen times that of its main rival. The reason, as historians like Jim Dyos have pointed out, was its decisive role in the growth of a centralized state and a national market.

Given such precedents, it is no surprise that doctrines of centrality have permeated disciplines dealing with the spatial distribution of people and things. In geography, for example, there is the highly influential central place theory of the German Walther Christaller and Jefferson's 'law of the primate city', with its insistence on the super-eminence, in national influence, of the largest city. Such models deal in ranking systems, hierarchies and pyramids – in other words, in authoritarian concepts. These models, in turn, permeate laws, culture and behaviour, casting a shadow even where the substance has long since vanished. Urban land markets, for instance, are bedevilled by what the geographer David Harvey calls the 'assumption of centricity' in which the abstract notion of a rental pyramid determines values and thus the use to which land can be put: a curious example of the tail wagging the dog.

Centrality also accentuates differences between city and country-side. In an influential series of essays in the 1950s, Robert Redfield and colleagues from Chicago University coined the term 'orthogenetic' to describe cities in the phase of so-called 'primary' urbanization in which urban elites – priesthoods, monarchies and bureaucracies – are seen as expressing and representing the old beliefs of the surrounding rural folk culture. In 'heterogenetic' cities of the secondary phase of urbanization, by contrast, the city assumes the role of service station while the hinterland becomes its food basket. Such cities, like the biblical Sodom and Gomorrah, are then characterized as wicked, or alien, breeding a distinctive and different 'urban way of life'.

Integral to their theories was the notion of a community of interest, or the lack of it, between city and country. The most interesting part of their discussion, however, is when they desert their exclusively cultural brief for a short examination of the sacred geography of ancient India, and in particular the spread of Hinduism. This, they point out, may well have been facilitated by its polytheistic beliefs, which made it easier to incorporate alien gods into the new system of worship. The same religious beliefs, however, determined the shape both of Indian cities and Indian villages. Each had a sacred centre in temple, tank and garden.

Here, strikingly, is one example where a decentralist metaphysic – in this case, polytheism – underpinned a sense of common allegiance between city and country, a sympathy which was all the stronger because it was embodied in the landscape and could be directly experienced as a part of daily life. It is also a convincing demonstration of how deep-seated social and cultural beliefs express themselves in patterns of land-use. As will be argued later, the relationship between belief, behaviour and land-use is crucial in the impact of environmentalism – another strongly decentralist philosophy and a powerful candidate for the role of late twentieth-century folk culture – on the city.

Community of interest was also provided by the physical presence of country in city. This could take several forms. At its simplest, it was the carrying on of agriculture, animal husbandry and viticulture within the city boundaries, often as an insurance against siege. A more complex form was that hinted at by Braudel in his striking description of European towns after their eleventh-century renaissance as the countryside 'revived and remodelled'. In the concept of the 'geomorphic', literally 'earth-shaped', city – in which the natural landscape, or more precisely, a feeling for and sense of the natural landscape, speaks through the built form – the balance of the two elements, earth-made and man-made, is more subtly achieved. And the best example of geomorphic design happens to be the most strikingly beautiful city in the world.

The grouping of 117 islets and shoals round the Rialto in a shallow and silty lagoon provided the blueprint for the creation of Venice as surely as any paper plan. But Venice's peculiar beauty – its disconcerting look of a city rising magically from the waves – lies also in the embellishment and elaboration of this pattern of islands by Byzantine, Renaissance and Baroque art. It thus deserves to rank as one of the finest examples of what the landscape architect Ian McHarg calls 'design with nature'. Yet if modern civil engineering technology had been available to the Venetian refugees when they fled the mainland and made their capital on the Rialto in AD 812 it is doubtful if its lacework of waterways would have survived in its present form. In that sense it is a marvellous accident.

In the broad sweep of post-Renaissance urban development, however, earth-shaped design has been eclipsed by the spread of man-made geometry. The orthogonal, or grid-plan, city was usually the simplest and speediest answer to the questions posed by conquest, colonization or industrialization: a sense of the local landscape and all that went with it – memories, allegiances and roots – was often a luxury which colonizers and industrialists simply could not allow their captive populations. But the final eviction of geomorphism from the modern city stemmed from the slow withering of relations between its people and its land. Farming, increasingly seen as unrefined and unhygienic – a response dating broadly from the eighteenth century – was exiled, delegated to the peasantry or cooped up, much diminished, in allotments and back yards. But land is a living presence as well as a productive resource. And while technology grew increasingly expert at obliterating topography, it was the complex change in attitudes to land marking the transition from feudalism to capitalism which lent, as it were, theoretical underpinning to the technology.

Land, formerly a gift, became a commodity. From being a dispensation, rarely to be alienated or forfeited, it became freely traded in the market. Ownership, once diffused and pluralistic, was privatized, and notions of exclusive property rights grew as those of stewardship diminished. A consequence was the proliferation of economic theories treating the city as a form of *tabula rasa* – what might be called 'pure space', devoid of characteristics other than those introduced by men. Robert Haig's 'frictionless' model for determining city layouts was among the most influential, forming a kind of blueprint for the twentieth-century business city.

The cumulative effect of these developments was to emphasize, almost to caricature, the alien nature of cities: they became, as it were, artificial islands in a natural sea, having little or nothing in common with the surrounding folk culture. Hence the fundamental importance of the rise of ecology. This is explored more fully in Chapter 4, but a few points must be noted here.

Ecology, like economics, is rooted in the Greek word *oikos*, meaning 'home' or 'household'. It not only offers insights from the study of plant and animal communities as a way of explaining the behaviour of human societies, but also speaks of affinities and analogies between the two. Moreover, because of its ability to comprehend the whole yet detail the parts – that marvellously sinuous concept of the ecosystem is a case in point – ecology offers a guide to the behaviour of individuals which is at the same time a guide to the wider setting – the ecological community – in which those individuals live. Economics, which has increasingly abandoned the wider moral and political framework canvassed by Smith, Ricardo and their contemporaries, has signally failed in this role, degenerating into an arithmetical accompaniment to the distension of human society. The immediate point, however, is the radical transformation in outlook resulting from the application of ecology to cities, whether as a tool of intellectual analysis, a technique of town planning or a critique of human behaviour.

In concepts like that of carrying capacity, ecology deals with the size and population of communities and their resources. It is implicit, for example, in observations like Arnold Toynbee's that the size of a city was dictated by how far a farmer and his horse could travel in and out on the same day and the numbers that could be supported within that radius. It is also implicit in questions of settlement growth. The Greek cities solved the problem of growing numbers by simply throwing off new colonies before they grew too large – usually well before their population reached 10,000, according to Mumford. Only one provincial city in England in the time of Henry VIII exceeded 10,000 people – this was Norwich, estimated at about 12,000 – while the average population of the 3,000 settlements in medieval Germany granted city status was a tiny 400. In the eighteenth and nineteenth centuries cities not only became spectacularly large; they also grew spectacularly fast. Population growth concentrating at a particular point produced disorder, stress and another source of ecological instability, the monoculture – in botanical terms, a pure stand of a single species, like the American cotton fields. The methods developed by the Victorian engineers and builders to cope with such numbers savoured of wartime: the concentration of resources and power into the hands of the few and hence the relative dependence of the rest. The burgeoning enthusiasm for alternative small-scale energy technologies in the 1970s was partly a response to the classic weakness of monocultures – their total vulnerability to a single source of attack. The nationwide strike was the culprit, capable of paralysing not only whole cities and technologies but an entire society. But monoculture was also reflected in the layout of the new industrial cities. Their key components – homes and workplaces – were standardized and then simply replicated in an apparently endless linear march over the landscape: the metaphor most favoured by the Victorians was that of disease, notably cancer with its overtones of disordered runaway growth.

Nineteenth-century urban growth deprived the mass of the new city-dwellers of regular contact with natural landscape on a scale without parallel in history, leaving the way open for the mythologizing of rural life which is so distinctive a feature of modern fiction. Of at least equal consequence was the syndrome identified by environmentalists as 'resource blindness', in which the food chain appears to shrink to a brief interlude between supermarket shelf and dustbin, leaving vast areas of vital activity shrouded in ignorance. Historically, resource blindness seems inseparable from large cities. Rome is its classic example, falling a victim ultimately to widespread soil erosion caused by overgrazing and deforestation, to the resulting chronic food shortages, and to the crippling taxes and military burdens imposed on a rural hinterland to fuel the squabbles of a metropolitan elite and the appetites of a teeming urban underclass.

To sum up, therefore, old urban traditions of centrality were fused with the demands of managing a mass society to create powerful new corporate cities, instruments of the concentration of political and technological power. Such cities were increasingly isolated from their hinterland, so that city and country became mutually exclusive realms. Yet when the prospects of civilization without cities were advanced in the later twentieth century, the old cleavages acted to block them. The conquest of space represented by new transport and information technologies – the car and the telephone, for example, and more recently the computer – offered immense scope for decentralization and for greater interchange between city and country. The changed realities of political and social power, and of religious belief, pointed in the same direction: towards pluralistic value systems in which the concentration of legitimacy at one point became increasingly unacceptable. Environmentalism, as we shall see, is playing a decisive role in this process. Instead, urbanism ran into an intellectual cul-de-sac, in which the only way forward appeared to lie in prolonging and magnifying the idea of the city. The result, ultimately, was a gross and peculiarly wilful distortion of reality, a state of mental transfixation slowly spreading throughout an entire culture in which escape from the city was ever more urgent and ever more unavailable. To a large extent we still inhabit the landscapes of that neurosis.

The key lies in elusive concepts of identity and, by extension, unity. By any criterion these are essential to the definition of a city. They explain why Max Weber says that a city must have walls and why Braudel argues that the open or unwalled town represents everywhere an earlier stage of culture than the closed town. They also explain why the dispersed city, lacking a coherent centre or focal point, mystifies and disappoints visitors. Los Angeles, spread-eagled by the motor car, is probably the best known. In each case the analogy is, broadly, organic: the city, the greatest physical expression of communal or cellular living in humans, must have a nucleus and a boundary. A minimum of differentiation from the outer world is intrinsic to its existence.

Such a conception of cities led instinctively to attempts to control their size. Resistance to the growth of London, for example, dates back at least to the time of Elizabeth I. By the late nineteenth century, the desire to contain cities had reached a new pitch of intensity, fuelled by their many horrors as places to live. A countervailing Utopian tradition was thus adumbrated, a self-conscious appeal to a smaller scale of communal life. In the most influential of the late Victorian model community programmes – Ebenezer Howard's garden cities – small, nucleated towns of 30,000 people were envisaged, each with a garden at its centre and five-sixths of its area reserved for farming. The town is divided into residential lots measuring 20 feet by 130 feet – enough, he calculates, to feed a family of five and a half people. Each town is separated from the next by two miles of open country and the ensemble of six garden cities and one central city together make up the 'social city' with a population of 250,000. But the pattern of social city could be repeated indefinitely – leading to a low-density, pastoral, polynuclear style of settlement focused on and interwoven with large swathes and patches of green.

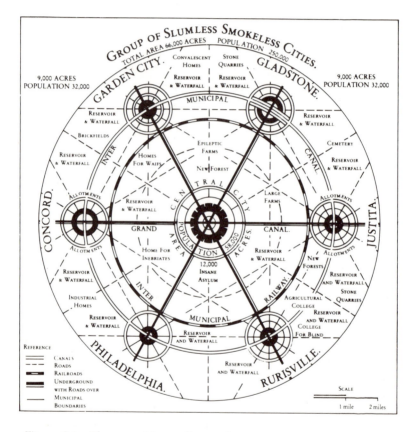

Figure 2.1   Ebenezer Howard's social city

Howard is, of course, cited as the father of the green belt. Yet the essence of his garden cities was the interpenetration of city and country. Instead, his swathes of green became *cordons sanitaires*, vaster and more intractable than any city wall. In 1974, for example, London's green belt was fifteen miles deep in places, increasing by 45 per cent in the next seven years until, by 1982, to the south-west of London in Surrey, it was well over twenty-five miles across. It had clearly become an instrument of containment – and, as such, an almost total perversion of Howard's ideas. In his vision of a marriage between town and country, he had decried the 'unholy, unnatural separation of society and nature'. Out of their union would spring 'a new hope, a new life, a new civilization'. The old orthodoxy was simply too strong for him.

The reason for the failure lay in the unfolding obsession with urbanism, in particular the apparently irresistible rise of the giant corporate city. The classic urban notion of organism, of a city-being with its own unity and identity, persisted to haunt a growth which had by now become cancerous, breaking all bounds of size, order and internal harmony. Any green field development of housing or industry was, as it were, a spore of this organism, a kind of secondary cancer: the land taken was instantaneously bounded, ruled and unfree. From Dickens's recoil, in *Dombey and Son*, at the 'giant's brick and mortar heel' of the city's outward march, to that remarkably influential tract, *Britain and the Beast*, published in 1937, to Ian Nairn's *Outrage* of 1955, the theme persists and intensifies. For Nairn it was no less than a prophecy of doom – 'the doom of an England reduced to universal Subtopia, a mean and middle state, neither town nor country ... [a] universal low-density mess'. The city, added Nairn, 'is today not so much a growing as a spreading thing, fanning out over the land surface in the shape of suburban sprawl.'

The emotional resonances touched off by this creeping and apparently inexorable corrosion of the countryside were increasingly striking. A new vocabulary had to be invented to convey the mystical, symbolic change of status represented by building development. Land 'disappeared' or 'vanished' under concrete – never, the implication was, to return. Countryside was 'desecrated' or, more generically and ambiguously, 'spoilt' – a term which touchingly hints at the inability of a society to understand its own psychic needs. But in two of the great classics of modern romance, *The Lord of the Rings* by J.R.R. Tolkien and *Watership Down* by Richard Adams, the connections become explicit. In both works, an animistic world of magic and mystery is specifically and starkly contrasted with the spread of development. The growth of the city, it seems, has become an affliction of the mind, from which an extravagant mental refuge has to be sought.

As part of the same fixation, the walls of the city have been enlarged until they embrace the whole planet. In Pierce Egan's *Life in London* in 1821, the settlement traditionally regarded as 'town' to the rest of England's 'country' was promoted to metropolis. In 1875 Greater London entered the vocabulary and in 1915, thanks to Patrick Geddes, conurbation. By 1961 the geographer Jean Gottmann was employing the term megalopolis to describe 'Boswash' in the United States – the 'almost continuous stretch of urban and suburban areas from southern New Hampshire to northern Virginia and from the Atlantic shore to the Appalachian foothills'. City-regions were already with us. 'Super metropolitan areas' also appeared in the 1960s. Later came 'super megalopolitan regions' – growth centres such as the Amsterdam-Paris-Milan 'golden triangle' or – even grander – the area bounded by Madrid, Rome and the northern Netherlands. Finally – in grandiosity, if not in chronology – came ecumenopolis, the world city glimpsed by the Greek planner C.A. Doxiadis. With an area of 2,770,408,000,000,000 square metres, ecumenopolis is, Doxiadis asserts, 343 times the size of megalopolis though only 2 per cent of the earth's land surface.

Boswash sounds unhappily like Nairn's universal low-density mess. Fortunately this turns out not to be the case. Flying over it, or moving through it, remarked the American environmental psychologist Amos Rapoport, makes one realize that 'surprisingly large areas consist of fields, forests and so on'. Similarly, Peter Hall and his colleagues, citing patterns of commuting, freight, post and rail travel as well as population densities, argue for the existence of a megalopolis England – a kind of vastly inflated Greater London – stretching from Sussex to Lancashire and from Dorset to Suffolk, but acknowledge that four-fifths of it, 'on any reasonable criterion', would be viewed as rural. Its justification, they argue, is that it represents the area of greatest pressures for urban growth.

In the megalopolis, the function of the city lives on but in nothing like its historical form. The conquest of physical space through technology – particularly information technology, the chief power-source of all urban elites – has both weakened the city's form – through actual physical decline and land vacancy – and established its civilizing function elsewhere. It is no longer a mark of 'urbanity' to be literate, to send articles by post or to travel by train. And since civilization no longer needs cities and people appear no longer to like them very much, their prospects seem dusty indeed.

Two contrasting viewpoints can thus be identified, each with its own far-reaching implications. In *Understanding Media*, Marshall McLuhan analysed the new forms of information technology and concluded, 'As electrically contracted, the globe is no more than a village.' Contrast this with 'the most complex urban form to appear in world history' – this is Hall describing megalopolis. Two quite different traditions are invoked. McLuhan chooses the village as his symbol of the computer and television age because he regards the new technology as decentralist in its tendency. As such it 'releases men from the mechanical and specialist servitude of the preceding machine age.' For McLuhan data-processing is the contemporary equivalent of the food-gathering phase of pre-urban nomadic man. Globally based, it must replace the city which 'as a form of major dimensions must inevitably dissolve like a fading shot in a movie'.

The true logic of megalopolis points in the same direction, to the abolition of boundaries between city and country in the face of forces which subsume and transcend them both. Misapplied, it is a ghost of a reality, serving merely to perpetuate the centralized and specialist living patterns, the top-heavy structures of power and thought and the oppressive communal life which have been the final fate of the city. And the fading of the city as a separate entity, the dispersal of its powers and population, may well serve as a preliminary to the eventual withering of nationhood, a concept that relies similarly on the principles of centrality, concentration and separation. In the urban world of the west, at least, such principles are falteringly but finally in retreat.

The city's resilience, however, should not be underestimated. Faced with the results of the 1981 census, the Social Science Research Council, as it was formerly called, embarked on a project to redefine the city. The argument was that the figures for urban population decline were too high: the true cities must be 'underbounded' by the census area. The old orthodoxies, it seems, die hard.

WILDERNESS, NATURE AND MUNICIPALITY

The crowning wrong that is wrought us of furnace and piston-rod lies in their annihilation of the steadfast mystery of the horizon, so that the imagination no longer begins to work at the point where vision ceases.

(Kenneth Grahame, Pagan Papers*, 1893)*

Civilization was finally achieved in the west in 1775 – fourteen years before the French Revolution and an event, conceivably, of longer-lived significance. The date is admittedly a little arbitrary. Three years earlier that fierce conservative Dr Samuel Johnson had resisted the pressure of the more fashion-conscious Boswell to include the word in his *Dictionary* instead of the traditional version, civility. Johnson's lesser-known contemporary and rival John Ash proved more open-minded, however. In his two-volume *New and Complete Dictionary of the English Language*, published in 1775, the word was set down as meaning the 'state of being civilized' as well as the act of civilizing.

Civilization, in other words, had come to represent a condition, not merely a process, of refinement – a condition which, as the nineteenth century progressed, involved increasingly elaborate ideas of arts, sciences and social order rather than merely *politesse*. It also took on a less savoury tone. During the same period in the eighteenth century – around 1760–70, according to Raymond Williams – attitudes to the city reached something of a watershed. Before this time, the city represented manners and refinement; after, it tended to be seen as oppressive.

The convergence was clearly not accidental. The change in attitudes to cities and civilization marked an important cultural revaluation. In many societies the act of human settlement was regarded as a form of benediction, a symbolic rescue from the wild. Mircea Eliade, for instance, speaks of settlement 'cosmicizing' nature by re-enacting the moment of creation, before which the material world is raw and formless. *Cosmos*, it should be noted, meant to the Greeks not only the universal order – an order which was in some sense alive – but also the form of the human body. In an important sense, settlement thus also represented the imprinting of human identity on nature – its incorporation into the landscape.

Even without this burden of symbolism, however, settlement, the most complete expression of which has remained the creation of cities, was capable throughout much of history of being seen as an unalloyed benefit – the communal extension of civilized life. The souring of attitudes to the city has been accompanied by a corresponding passion for its opposite: those beliefs most easily summed up under the heading of 'wilderness'. Hence the conclusion of that diligent student of man's relationships with his surroundings, Y.F. Tuan. The last century or two, Tuan argues, has seen an exact inversion of long-established values. Wilderness has become sacred and settlements profane.

A deep-seated change has clearly occurred. Whether it represents such a complete break with the past is another matter. Wilderness has fuelled emotions and fired the imagination from the earliest times. Its forms have varied, however – and a measure of this variation is the extent to which previous attempts to implant the wild in cities are now judged to have failed. Chief among these are the Victorian and early twentieth-century parks introduced in Britain and America by private and municipal benefaction, usually as an afterthought, to make city living tolerable.

The genesis of these parks is illuminating. The eighteenth century, by and large, disliked wildness. 'Wild Nature,' declared Count Buffon, probably the most influential naturalist of the period, 'is hideous.' Defoe called the Lake District 'barren and frightful' and Johnson was similarly uncomplimentary about the Scottish Highlands – 'matter incapable of form or usefulness, dismissed by nature from her care' was his judgment.

It was, nevertheless, the created landscape of the later eighteenth century – that vista of rolling parkland, a lake and clumped or dotted trees associated with the names of William Kent, Humphry Repton and, above all, Lancelot 'Capability' Brown – that provided both the inspiration and the basic form for the urban parks that followed.

The Brown 'style', developed at Lord Cobham's country seat at Stowe in Buckinghamshire after 1741 and later applied at Syon, Chatsworth and Longleat, dominated the design of aristocratic arcadias long after his death in 1783 – an ascendancy summed up by his appointment as Master Gardener to George III. It gradually assumed the status of an international landscape language, copied throughout Europe and the United States. It remains many people's idea of what leisured countryside should look like. It was thus natural that the Victorians should turn to it when they came to consider the human consequences of spreading cities and spiralling densities.

J.C. Loudon, that assiduous promulgator of vegetative exotica and advocate of herbaceous borders and the 'gardenesque' style, was the first modern garden designer to advocate publicly owned parks: his first article on the subject was in 1803. In 1840 they were recommended by the Select Committee on the health of towns, which saw them as a thinly disguised instrument of crowd control. Parks would, it was argued, improve morals, reduce disease and crime, and, as one campaigning MP expressed it, prevent discontent 'which in turn led to attacks upon the Government'.

People crowded into the new parks with a desperate enthusiasm, poignant testimony to a wider sense of deprivation. In Victoria Park, the first great public park in London, 25,000 bathers plunged into the open-air lakes before 8 a.m. On one June Sunday shortly after it opened 118,000 were estimated to have visited it. Sheer numbers threatened to overwhelm them and the authorities responded by installing railings, banning games and employing park police to break them up, forbidding contact with grassed areas and, later, replacing the grass with paving or gravel which could better survive the tramp of so many feet.

The parks that sprang up in such numbers in the 1840s and after – relatively few of them, unfortunately, in the areas of highest density and greatest need – were in many cases the creatures of the new municipal corporations, set up as a result of the local government reform of 1835. There was thus a whiff of populism about them: some, as in Halifax and Hull, were called 'people's parks'. The reality, however, was increasingly paternalistic: a well-drilled multitude going through its recreational paces. For the first time in history on such a scale, escape from the city had been identified with a vision of natural countryside and the vision had been given a physical presence inside the city. It was, in its way, a great adventure, but it was an adventure which by the end of the nineteenth century had become markedly bureaucratized.

A clear sign of this came in the administrative designation of parks as 'open space', a concept clearly paralleling that of 'frictionless space' in location economics: both imply land deprived even of topographical content. The first Open Space Act came in 1887 and in 1925 the National Playing Fields Association laid down its still unrealized target of ten acres of open space for each population unit of 1,000. A tenth of this was to be 'ornamental' and the rest 'active'. Open space, it should be stressed, is not merely a useful means of measurement or a term of administrative convenience. It is a way of seeing and thus of behaving. A man who surveys a rich vista of hills, woods and water and can see only open space suffers a form of dyslexia, a disablement of vision comparable to describing a symphony as a collection of quavers and crotchets or a sunset as an atmospheric phenomenon. That society or government is responsible for this designation makes little difference: the blind spot has merely become collective. Such a society will then produce landscapes which are little more, literally, than open space: bland, blank and monotonous. In a similar way the designation of farmland as 'open background' from the 1940s onwards was reflected in blank spaces on the maps used by town and country planners and blank spaces in the minds of the urban public. As we shall see later, this had an enormous effect on the countryside.

If Brown was the grandfather of the Victorian park, its immediate paternity was shared between Loudon and Sir Joseph Paxton, the apprentice gardener who became the architect of the Great Exhibition. The winding drives and artificial hills incorporated by Paxton in his epoch-making design for Birkenhead Park in 1843 represented the application of Brown to the cities. Loudon, meanwhile, declared that no modern garden, of the kind required by the new villa-owners and suburbanites who viewed Brown's country-house patrons as their rightful ancestry, could be considered tasteful without an array of foreign or 'improved' native shrubs and trees. A style thus evolved of sweeping lawns and ornamental trees punctuated by self-consciously horticultural displays. It was a marriage of the aristocratic and the bourgeois, the country house and the suburban villa. It remains the pattern of most urban parks today.

Certain lessons and paradoxes emerge from this brief history. Perhaps the chief paradox is that an age which could not abide the wild fashioned the landscapes of an age which came passionately to crave for it. Another is that a style which owed its success to naturalism – Brown's landscapes were positively Spartan compared with the statue-haunted arabesques of rival schools – could become so laden with artefact as to merit, finally, the label 'ornamental'. A third, less paradoxical but certainly remarkable, is the bureaucratiz-ation of a widely accessible and highly popular form of aesthetic expression – landscape design and appreciation.

The eighteenth century, it should be remembered, invented landscaping – as opposed to gardening – and regarded it as an art. Alexander Pope declared that 'all gardening is landscape painting' while Pope's younger contemporary Horace Walpole remarked in his *History of the Modern Taste in Gardening* that 'Poetry, Painting & Gardening, or the Sciences of Landscape' were the 'Three New Graces who dress and adorn Nature'. Subjecting private delight in natural beauty to the edict of the municipal parks department was rather like nationalizing the composition of string quartets.

If the act of human settlement 'cosmicized' nature, effecting a radical and enduring change of status, the act of 'unsettlement' involved in the import of country into city proved far more difficult. The deliberate unmaking of civilization was something of an historical hurdle, not to be cleared at the first attempt. The parks, fragile enclaves of countryside, could not maintain their integrity in the face of physical and cultural pressures. They became adjuncts of the city, an extension of its planned and constructed fabric: hence the emphasis on the ornamental, the architectural and the sculptural, all aspects of the city as artefact. In an important sense, their shape, size and position – most notably the fact that they were physically cut off from the 'true' countryside – dictated their regime. So, too, did the collectivist mentality which went with them. This stressed standardization of treatment, conspicuous provision – a proof of value for money – and, as a corollary, active design. Parks thus became entertainment, placing the municipal marigolds firmly in an urban tradition stretching back to the *panem et circenses* of imperial Rome.

One of the most perceptive comments on the modern municipal parklands has come from the Swedish landscape architect Roland Gustavsson, who speaks of their being designs for 'sunny days': there is little thought for days of wind, cloud and rain, for that peculiar sense of privacy and comfort which comes from standing in a dripping wood and feeling oneself to be in a closed outdoor room – what Gustavsson calls an '*innerstand*'. But the important point is not so much the nature of the design as the fact of the design itself. The American sociologist Herbert Gans argues that one of the main causes of the failure of urban landscapes has been 'the designer's preoccupation with aesthetics'. Ted Relph uses the term 'rational landscapes' to describe how progressive humanism has this century so extended the ideals of comfort and convenience to living space that the result everywhere has been the 'casual eradication' of local distinctiveness. Relph calls this placelessness. Allan Ruff and Robert

Tregay, advocates of the ecological school of landscaping, talk of 'value-laden landscapes' which pre-empt a personal response. In any created landscape, clearly, there has to be design. The issue is whether it should obtrude to the point of self-advertisement. In the history of aesthetics, as Horace's adage *ars est celare artem* reminds us, it is not a new problem.

The picture thus emerges by the second half of the twentieth century, in the immediate post-war decades, of an urban landscape tamed and tidied, its parks ruled off into trim, brightly floral segments and regulated by council edict. Grassland is manicured – the lawnmower was invented by a Victorian textile designer – trees are specimens and shrubberies exotic. Park gates are closed at dusk, ball games are forbidden and active use – as opposed to passive spectatorship – generally discouraged. Vandalism is rife, both inside the park and in the tower blocks that surround it.

It is a sad end to a movement which, in its aims and its achievements, represents one of Britain's most distinctive and influential contributions to cultural history. Significantly, criticisms of the modern cityscape and its parklands provide remarkable echoes of those voiced three centuries ago by the originators of the British landscape movement. The focus of protest then was what Shaftesbury called the 'formal mockery of princely gardens' – the prevailing French and Dutch geometric style embodied most resplendently in Andre Le Nôtre's laying-out of Versailles from 1661. It was, as Shaftesbury's words imply, a self-conscious gesture of Anglo-Saxon libertarianism – a revolt against the Continental autocracy typified by Versailles – and among its main technical innovations was the popularization of the ha-ha, or sunken ditch.

The spread of the ha-ha is an event of underestimated importance in the story of men's attitudes to nature. In modern history, it forms an important stage in the shift of emphasis from the near-at-hand to the distant and in the increasing inability of human beings to find emotional significance in bounded space. Hence the progression from the medieval walled garden, with its sense of privileged enclosure, to the American national park, where the emphasis is specifically on the absence of walls.

The ha-ha – originally 'ah-ah', an expression of surprise and delight – was mentioned in a French gardening treatise of 1709, arrived in an English translation in 1712 and subsequently became a favourite device of Brown's. Among its earliest users, however, was Charles Bridgeman, the designer of Kensington Gardens in London.

In 1730, when Bridgeman was at work there, Hyde Park beyond was considered to be countryside. By sinking a ha-ha at the perimeter of the new gardens, an invisible boundary was created. Stray cattle were kept out but there was no wall or fence to disrupt the union of garden and landscape, of the domestic plot and the distant skyline. Illusion made settlement and nature a continuum. Nature was restored to man.

Pope described the aim of the landscape movement as 'calling in the country', but there seems in this new obsession with distant vistas, and in the rejection of the garden's historic function as charmed enclosure, something even more profound at work – not merely the cloying of civilization but the emerging outlines of a fixation with unexplored frontiers and far horizons.

But perhaps the most important feature of the landscape movement was its fusion of ideals of political and psychological freedom in a view of nature which has, quite literally, refashioned the world in which we live. One of its abiding lessons is therefore that landscapes altered by man – 'cultural' landscapes, as they have been called – express political and social judgments which can age rapidly. Trees, rocks and bodies of water are usually slower to change. Landscapes can thus get hopelessly out-of-date. And if the sense of suffocation and sterility experienced by many when viewing the modern city landscape marked the failure of the first campaign to 'call in the country', a new campaign was clearly called for.

The picture of modern cityscapes painted above is, like all composites, a little unfair. Small pieces of countryside were, for one reason or another, passed by in the city's distension and simply forgotten, invisible to the bureaucrat and familiar only to a few local people who were able to rediscover there some of the medieval delights of enclosure. Larger areas further from high living densities retained some of their original character – though Steen Eiler Rasmussen's remarks on one of the most prized of these, Hampstead Heath in London, is mordantly apposite. The average Londoner, said Rasmussen, writing in the 1930s, is astonished to hear the heath described as a park, since he lives 'in the happy delusion that it is a no-man's land where everybody can do as he likes'.

But some of the stoutest resistance to civic embellishment came in those parklands that were not gobbled whole by an amoeboid city but maintained, as it were, a common front with the countryside, in the style of Ebenezer Howard's social city. For all central London's heritage of royal parks, leafy squares and gardens, Britain's park-builders achieved nothing compared with the park systems designed in America by the heirs of Capability Brown.

Probably the most impressive of these was that designed in Boston with the guidance of Frederick Law Olmsted, the world's first self-styled landscape architect and the chief author of Central Park, New York. Starting in 1875, two concentric rings of parkland were laid out, the inner with over 100 miles of walks and rides, the outer linking the forested heights of Blue Hill and Middlesex Fells with beaches, harbour and the Charles River. Linking the gardens at the centre of Boston to the outer ring was Commonwealth Avenue, 240 feet wide and with four rows of trees shading its central mall. Other American cities shared in what became known as the City Beautiful movement – by the 1870s, Chicago was already known as the Garden City – but Boston remains its greatest achievement.

Olmsted had visited Britain in the 1850s and declared that America had nothing to compare with Birkenhead Park. Half a century later, the position was reversed. The Americans had taken a decisive lead in the design of green city landscapes. Whereas in Britain the movement had all but petered out, regressing to garden design, the Americans in 1899 founded a national professional body and the following year set up the world's first university course in landscape architecture at Harvard. A powerful reason for this American advance was their collective experience of the wild.

The civilized and the primitive have never been brought face to face quite so dramatically and on such a large scale as in the colonization of the North American continent. Out of the encounter came the western world's first self-conscious cultivation of wilderness – an odd phrase testifying to a remarkable cast of mind – in a line which stretched from the Leatherstocking novels of Fenimore Cooper, the first written in 1823, through the transcendentalist nature worship of Thoreau and Emerson, to its expression in public policy: the preservation of Yosemite and Yellowstone National Parks in 1864 and 1872, and the spread of the City Beautiful movement. Thoreau, declaring that 'all good things are wild and free', proposed wilderness areas of between 500 and 1,000 acres for every city. Olmsted, echoing this, believed parks helped to prevent neuroses and 'vital exhaustion' and wanted to see patches of wild forest preserved close to cities: an aim that was near to being realized in Boston. But his principles were expressed most popularly in his axiom that 'in the park, the city is not supposed to exist' – as good a comment as any on the ideals of the nineteenth-century park designers but one which now has a distinctly forlorn sound.

The modern appetite for wilderness may have had its roots in Europe – Rousseau's sentimentalized 'noble savage', the rumblings of Wordsworth and the Lakes poets, the growing fondness for the 'sublime' qualities of mountain and moorland – but its most vigorous growth was in the United States. Not only did North America represent a purer version of the wild than Europe – the plains vaster, the mountains more immense, the forests thicker and less peopled – but the Americans themselves, particularly in this century, lived more affluent and air-conditioned lives. Hence the development of a wilderness industry: backpacking or canoeing – by ticket only – in the Grand Canyon or down the Colorado River, or simply sampling the delights of Yellowstone safe in the confines of a 'wilderness experience booth'. Demand is immense. In 1963–64 forty-four people made the Colorado River trip by canoe or raft. By 1972, when a ceiling on numbers was fixed, there were 16,432 – 373 times as many. In 1981 would-be canoeists faced an eight-year waiting list. Yet 91 per cent still considered it to be genuine wilderness.

The trouble with explanations which trace a demand for the 'natural' to an excess of the 'artificial', or which simply contrast wilderness with civilization, is that they do not really explain. Dr Johnson would have considered himself highly civilized, but would have viewed sleeping rough under the stars as a mark of penury or extreme folly. Nor do the pronouncements of the wilderness lobby help noticeably. There is Thoreau, for example, declaring grandly, 'In wildness is the preservation of the world.' Or there is that ambivalent Jesuit poet Gerard Manley Hopkins proclaiming in 1881, 'Long live the weeds and the wilderness yet.' A century on, Frank Fraser Darling argues in the more laboured style of contemporary environmentalism that 'natural wilderness' is a factor for world stability, not some 'remote and inimical place'. Its co-operation with us, he adds, is 'unconscious'. At the end of each such declaration – and there are many more – a single question resonates: why?

It is evidently an important question because the wilderness urge is clearly the latest and most extreme version of a long-standing and apparently deep-rooted nature quest which may, if it is indeed a generic ill of civilization, come to pose serious threats, physical and psychological, to an increasingly crowded and affluent planet. Growing millions inside America and Europe, and many more outside them, may come to crave the Colorado River experience and what it represents – the challenge of survival in a solitary world.

Wilderness itself is thus jeopardized or, like the Victorian parks, it becomes vastly oversubscribed, a meagre resource parcelled out by well-meaning bureaucrats, subjected to rules and management regimes which render it valueless. Or, worst of all perhaps, it becomes a gigantic illusion, a landscape devoted to play, devoid otherwise of meaning, like a vast theme park in which life or death situations are merely simulated and experience entirely disjoined from its consequences.

That is one possibility. Another, offering more ground for optimism, is that the search which led out of the walled garden, over the ha-ha and into the far horizons, may, conceivably and ultimately, lead back home again.

4

THE RECOVERY OF THE PRIMITIVE – ENERGY, ECOLOGY AND GOD

They found a delicacy for dinner, a certain shelly, slow-moving water animal very good to eat .... 'You eat them, Yahan, I can't shell something that might speak to me' [Rocannon] said, wrathful with hunger, and came to sit beside Kyo.
Kyo smiled, rubbing his sore shoulders. 'If all things could be heard speaking ....'
'I for one would starve.'

(From Rocannon's World, *Ursula K. Le Guin, 1966)*

'We have created the word monergy to expose how a waste of energy is a waste of money. It's a word that will soon be in the Oxford Dictionary.'

(Peter Walker, Secretary of State for Energy, speaking at the National Exhibition Centre, Birmingham, on the launch of the Government energy efficiency campaign, 12 November 1985)

'This is a spiritual matter. He wants to go to his place of worship.'

(Counsel for David Clapson, aged 34, a Portsmouth antique fire-grate specialist, who unsuccessfully sought an injunction at Winchester Crown Court to gain uninterrupted passage to Stonehenge for the 1985 summer solstice)

If, in the later twentieth century, the country was once more to be called in to restore and recreate human settlements, much would be found to have changed. Not only have cities spread and a wilderness cult arisen but a vision of the natural world has taken shape, its roots in history but its overall perspective fresh, vital and highly – often disconcertingly – distinctive. It is this vision, or elements of it, which since the 1960s has fuelled the movement known as environmentalism, and this vision which is reshaping our cities and our ideas of what constitutes the human home, or *oikos*.

Three and a half million people belong to environmental organizations in Britain, a fast-rising membership total double that of the three main political parties and three times the (declining and overestimated) numbers of Anglican Sunday churchgoers. The American conservative philosopher Robert Nisbet has described environmentalism as the third great wave of redemptive struggle in western history, after Christianity and Marxism. But perhaps the best testimony to its influence is the slow colouring of an entire culture with its imagery and attitudes, from natural childbirth to herbal shampoos, from the animal rights movement to the advertising industry – the unquestioned association between the natural, the pure and the good assumed by the latter ranking as perhaps the ultimate cultural accolade.

In any gathering of 'greens', for the anthropologically-inclined observer, shadows of ancient beliefs abound. There is totemism, for instance – the identification of men with plants or animals. Emblems of whale, seal and panda are ubiquitous: through them man claims kinship with animals. There is also autochthony – people's sense, as Mircea Eliade describes it, that they belong to a particular place, which they express with a sense of 'cosmic relatedness deeper than that of familial and ancestral solidarity'. Autochthonic beliefs, associated with the processive symbolism of the cavern in the religion of palaeolithic hunter-gatherers and later with the worship of the Earth Mother, led the Romans to call a bastard *terrae filius* – son of the earth – and to the widespread folk custom of giving birth on the soil: hence the expression 'native' soil. They also clearly share some form of psychological ancestry with the growing assertion of local, regional and ethnic identities, the search for roots and a sense of 'belonging'. Bioregionalism, for example – the idea now approaching in the United States the status of an overtly political movement – holds that units of local or regional government should reflect zones of physical relief and plant and animal communities: so-called 'ecotopes'. Geology and, more importantly, ecology, are grafted on to autochthony.

The ancient beliefs are not necessarily swallowed whole, however. What the evidence indicates is a deliberate, often highly cerebral, recovery and reformulation of the primitive: a search for, and rediscovery of, a set of symbols which correspond to some deeply felt but poorly understood personal need. In effect, it is the twentieth-century consumer ranging at will through a vast psychological supermarket, its shelves loaded with the cultures of millennia. It is this element of choice which distinguishes the recovered primitive from the genuine superstition, and which also explains the unease felt by a predominantly rationalist and reductionist age.

In one specialized sense, of course, the old beliefs do survive. The various forms of recorded fact and fantasy associated with a literate, urban society – the novel, the film and the encyclopaedia – represent a storehouse of symbols theoretically far richer and capable of far greater subtlety of expression than the single-channelled folk tradition of an oral culture. What is lost in the empathy of immediate 'unconscious' assent may be outweighed by a more comprehensive and multi-layered experience.

What the green mythologies seek to achieve, and what distinguishes them clearly from the undergrowth of cults which has followed the slow collapse of organized religion, is a new union specifically between man and nature through the integration of two increasingly divergent traditions: on the one hand, the rationalist and technological, and on the other, the imaginative and the primitive. By convention, of course, these are the preserves of city and country, separate realms whose growing physical differentiation has accentuated the psychological divide. More will be said of this divide later. The immediate point is that this fusion reaches its most complete and coherent form in ecology.

Some definitions and qualifications are necessary here. The term 'mythology' is not used disparagingly but to describe a system of belief which seeks to explain the universe both intellectually and imaginatively – to present a truth which engages the whole person, not merely the abstract side represented by the neo-cortex. This is the older meaning of myth: its newer meaning, implying fiction or superstition, is a product of the mind-body duality associated with the French philosopher René Descartes and the revolution in modes of thought which this signalled.

The Cartesian approach prompted a revaluation of man's relationship with the natural world, as discussed later in this chapter. Its emphasis on abstract, analytical, mechanistic thinking, its assertion that 'there is nothing included in the concept of body that belongs to the mind; and nothing in that of mind that belongs to the body', has also led in the four centuries since he wrote to an acceptance by western culture of the 'scientific method' so total and unquestioning that it has become an unconditional reflex – a form of contemporary dogma. Not only does suspicion immediately and instinctively fix on whatever cannot be reduced to its component parts and put under a microscope; the attempt to use scientific knowledge to explain perceptions of imaginative or emotional truth is doomed. This is not, says the spirit of the age, what scientific knowledge is *for*.

The business of living is thus subjected to a massive and traumatic divide between what is 'real' and what is 'irrational'. The new mythologies seek to heal this great cultural wound western society has inflicted on itself. To do this they must speak the language of science because it is the universal language. They must also invoke its metaphors because it is through those metaphors – camouflaged imprecisions of thought – that science touches most nearly on matters it cannot explain and in those metaphors that western society comes closest to an imaginative truth along its chosen rationalist path. In effect, the metaphors act as Trojan horses through which body can re-enter mind. It is thus no coincidence that they abound at the frontiers of science, from the genetic codes of molecular chemistry and the quarks and charm of particle physics to the black holes and anti-matter of astronomy.

A second qualification concerns describing the new mythologies as 'green'. It is probably more accurate to say that they culminated in environmentalism as the best and readiest expression of a mood, an increasingly powerful cultural undertow. For what the mythologies share is not only sympathy with an identifiable 'counter-culture', but the conviction that only through recovery of a personal primitive – whether magical or severely practical – could a start be made on rejoining mind and body and forging a new vision of reality, at once intellectual and spiritual. The bond uniting the disjointed and heterogeneous coalition of concerns that makes up the wider green movement is its devotion to ecology.

If environmentalism is the twentieth century's chosen path to salvation, its sacred text is *Silent Spring*, Rachel Carson's indictment of pesticide use published in 1962, and its 'prelapsarian idyll' is the American continent before the advent of the Europeans. This was Nisbet's formulation and it neatly captures the spiritual drive behind the green movement. It is this quest for the spiritual which has taken hold of ecology, once a well-scrubbed natural science with merely academic pretensions, and transformed it into a morality and a metaphysic.

As the second half of this book seeks to demonstrate, it is this metaphysic, rather than any narrowly conceived 'environmentalist' movement, which is now changing the form and function of our cities. George Sessions has made the useful distinction between 'shallow' and 'deep' ecology – the former emerging as a concern for environmental management and protection, the latter as a mode of perceiving the universe, its meaning and relationships. For a true picture of deep ecology, its persecution, defeat and final slow renaissance, one must therefore look at man's historical relationship with animals and plants. One must also examine how this has been imaginatively expressed since it is in his art and fiction, rather than in works of overt self-analysis, that the outlines of a deep ecology will become clear.

This is particularly true of cult fiction, notably that series of modern anthropomorphic classics which seems to touch such a sensitive cultural nerve. Anthropomorphism – the attribution of human characteristics to the non-human world – is a staple of literature, particularly that written for children, but in Kenneth Grahame's *The Wind in the Willows* (1908) and A.A. Milne's *Winnie the Pooh* books (1924–28) it took on a new and evidently compelling significance for adults – centred chiefly, it seems, on its association with a lost rural world of childhood.

In *The Wind in the Willows*, the deliberate invocation of the mythic and the primitive had already brought a brief and benevolent incarnation of the god Pan, but it is not until much later, in identifiably modern anthropomorphic classics like Tolkien's *The Hobbit* (1937) and *The Lord of the Rings* (1954), and Richard Adams's *Watership Down* (1972), that this element becomes so pronounced. All three elaborate the interconnectedness of the non-human world, imply it is of at least equal value to the human and invest the magical and the mythic with a new breadth and intensity of vision. In *Watership Down*, a saga of rabbits driven from their home by urban development, trances, visions and prophecies abound. In *The Lord of the Rings* men are merely one, and probably the least interesting, of several species of creature engaging in the ritual moral struggle of epic romance: sorcery and the paranormal are again commonplace. All these books are among the most read and treasured of the twentieth century: *Watership Down*, for example, is already the second-best selling Penguin on record. For Tolkien, a student of middle English romance and professor of Anglo-Saxon at Oxford, this was indeed a recovery of the primitive.

But anthropomorphism has been extended in *The Lord of the Rings* from animals to plants, and, it seems, to entire landscapes. Forests open, close, regroup themselves. Hills, fog and standing stones conspire together against travellers. The force which energizes them hovers on the brink of impersonality, a vitalism part vegetative and part magical. In the end, however, the portrayal remains – by an extremely thin margin – anthropomorphic because a recognizably personal entity is implicated. The spirit or 'barrow wight' that traps travellers in a burial tumulus is revealed by a 'long trailing shriek', when vanquished, to have a discrete identity. The old forest, though itself some form of entity, is capable of individuating itself as a race of 'ents', ancient and gnarled beings, but beings nonetheless.

Anthropomorphism is a simplified and tamed version of relationships thought to exist in *illud tempus*, the 'sacred time' at the beginning of creation which myth describes and re-enacts. Participants in a ritual re-enter that time and are thus 'reborn'. The myths of Paradise or the Golden Age – both universal, according to Eliade – contain several features in common: the immense and self-yielding fruitfulness of the earth; the resulting absence of work; and friendship with animals and knowledge of their language. Hence the imitation by shamans of animal cries – an act which gives them access to a 'spiritual life much richer than the merely human life of ordinary mortals'.

These beliefs can be found throughout ancient cultures. According to the animistic conceptions of the South African Bushmen and the Australian Aborigines, for example, all living things, particularly animals, were once people. And the 20,000-year-old cave paintings of south-west France and northern Spain, according to Gertrude Levy, were not 'art for art's sake' but a 'participation in the splendour of the beasts which was of the nature of religion itself'. The animals represented a 'vision of achievement', an already perfected species. Hence men revered them.

Wonder and veneration, or a simple sense of kinship, could survive in medieval and early modern society in the west as folk tales – of robins burying the dead with moss or of hares changing sex and sleeping with an eye open. But such attitudes were slowly driven underground by the ideologies that came to dominate western thought. The major tradition of Christianity took its cue from God's instructions to man in the first book of Genesis, reaffirmed later in the eighth Psalm, to subdue the earth and assume 'dominion . . . over every living thing'. Nature worship stood condemned as idolatry. Descartes, meanwhile, viewed animals as 'automata' – machines incapable, unlike men, of reason.

The pattern that emerges, of man intellectually and physically domineering over a subjugated and devitalized Nature, has probably been exaggerated. Alternative traditions remained, the 'minority' Christian notion of good stewardship, for example. Often the text seems to have been chosen to suit the need of the moment, which, in practical terms, increasingly involved the colonization and exploitation of new lands. But the Cartesian spirit also found expression in the classification devices adopted in biology and botany. These culminated in the binomial system of Linnaeus, beginning with the *Genera Plantarum* of 1737, which now forms the basis of classification in the life sciences.

By the time of Linnaeus conquest, colonization and resulting trade were producing plant discoveries in abundance, and there was a pressing need to impose order on botanical chaos. His system was an instrument of dominion in the best Cartesian tradition, and it was reflected in the enormous popularity of botanic gardens, of which there were some 1,600 in Europe by the end of the eighteenth century, according to a contemporary estimate by the Swiss botanist

Jean Gesner. These plant zoos, the specimens neatly labelled and pigeonholed, set a precedent for both the dusty glass cases of the great Victorian museums and the nineteenth-century mania for collecting – flowers for painting and pressing, rocks and minerals, butterflies, ferns and fish. Entombment or incarceration – the transfixed butterfly, the aquarium (invented in 1850), the stuffed bird – figured prominently. In abstracting creatures from their natural habitat, in reducing their identity to a binary function of genus and species and in immuring it in Latin, such attitudes narrowed and simplified, rendered one-dimensional, the older and richer sympathies between men and other creatures. The spectator-specimen relationship, through a glass mustily, reinforced the privation. Legends of robins and hares were unscientific and 'superstitious' – a word first used in its generalized modern sense, meaning irrational, in 1794. Anthropomorphism was banished to fiction.

The re-emergence of the older sympathies comes in a form distinctive to the late twentieth century. Evolutionary theory, amplified enormously by ecology, has plucked plants and animals out of their cages, where they stood as changeless representatives of a type, and placed them back in a particular living community. In this community, and only in this community, according to ecology, can the creature's full and proper identity be experienced. Moreover, it is an identity which responds to the minutest change in surroundings. From ecology old autochthonic myths receive sustenance. Localism is vindicated.

Ecology has also breathed new life into the minority Christian tradition of man as God's steward on earth, charged with the task of managing creation fruitfully. Its lesson is that of inter-relationship, dependency, and hence partnership. In the concept of the planetary ecosystem, fuelled by the sun's annual output of 1.73 billion megawatts and with a biomass of 1,855 billion dry tons – a figure which represents the plant life of forest, sea and grassland – the earth becomes no longer a human autocracy, but a federation of man, beast and plant, a vast co-operative in which energy is the sole currency of value, passing through various stages of decomposition, recycling and consumption to build flesh and cellular fibre alike.

The development of energy as a central cultural metaphor demonstrates perhaps most clearly of all how ecology has acted as the catalyst in the emergence of a new vision of the world with far-reaching implications. It is a vision which is scientific and yet also imaginative, spiritual and ethical. It is shared, moreover, by large numbers of people.

It is also a re-emergence of an older pagan world-view. The principle can be found, with a thin intellectual veneer, in the *logos spermatikos* of the Stoics – the vital heat, as Balbus expresses it in Cicero's *De Natura Deorum*, that permeates and sustains the world. It can be found in the Tao, the living principle which pervades the universe and from which sprang the 'superstition' of geomantics. Based on the principle of an active correspondence between human well-being and the earth forces that display themselves in landscape – so that men design their settlements 'to co-operate and harmonize with the local currents of the cosmic breath' – geomantics is now undergoing a notable renaissance. In Hongkong western planners have learned to accept the Cantonese view that a particular wooded hill or reed-fringed lake is good *feng-shui* – healthy or beneficent. This is partly because geomantics makes good aesthetic sense. But empirical studies have shown how landscapes affect stress in different ways. Researchers at the University of Delaware showed slides of rural and urban scenes to students, tested their anxiety ratings and also measured their brain alpha rhythms using EEG. They found that students shown 'natural' scenes, even undramatic pictures of thicket and hedgerow, felt safer, friendlier, more content. Students shown town scenes, even those chosen deliberately for their cleanliness and tranquillity, felt worse – sadder and more depressed, and also more aggressive.

Later tests showed, more interestingly, how natural scenes which featured water satisfied and contented people more than those which did not: cityscapes still fared worst. More recently, analysis of hospital records in Pennsylvania indicated that post-operative patients appeared to recover more quickly, with fewer complications and substantially less need for drugs – drug intake was reduced by almost two-thirds – if they could see trees through their window rather than a blank brick wall.

New research in disciplines like biometeorology and geochemistry, dealing with the impact of weather or soil on bodily health, points in the same direction. 'Sunshine' therapy – simulating the arrival of spring – has been used to treat patients suffering from depression. Negatively charged air particles, or ions, linked with improved mental and physical, notably respiratory, function and now capable of being produced at home or in offices through commercially available ionizers, are produced by running or breaking water: hence, perhaps, the attraction of spas, waterfalls, streams and seaside. All these theories would find a ready echo in that classic treatise on environmental determinism, Hippocrates' *Airs, Waters and Places*, which saw men not as cut off from the rest of creation by virtue of their consciousness – broadly, the scientific and the Judaeo-Christian view – but rather as vital units in a vital world, of a piece with that world and vulnerable to its moods and influences. The Hippocratic outlook is close to the view of 'deep' ecology. So, too, is geomantics. Moreover, for over two millennia western civilization had a ready means of explaining how man and the universe, microcosm and macrocosm, swam in the same living stream. This was the theory of the four humours, Hippocratic in origin, which through an elaborate series of correspondences – blood and air, bile and fire – portrayed men and matter as composed of the same physical stuff. In discrediting the theory of the humours, modern science also destroyed this easy dialogue between man and nature.

Stoicism and the Tao are 'civilized' versions of this vital principle. The much-debated mana of the Pacific Islanders, and its analogues in many other cultures, constitute a more 'primitive' form. Mana is perhaps best described as an indwelling force permeating landscapes, plants and living creatures, fraught with mystery and strangeness, associated with taboos and rites of passage, with birth, fertility and death. Since it is charged with such significance, it is easily conceived of as sacred. It was once understood as characteristic of that stage in the earliest development of religion which precedes animism. Animistic beliefs credit trees, rocks and animals with their own souls: the indwelling force, as it were, has sprouted personality.

Searching for an adequate definition of mana and its analogues, Mircea Eliade selects the notion of 'reality'. The primary vision of the North American Indians, he says, was not that the force was marvellous or extraordinary but that it appeared to represent 'real existence'. His conclusion recalls that of Gertrude Levy on palaeolithic religious art. For Crô-Magnon man, the animals were not merely beautiful but complete: they were an integral and timeless part of creation, utterly at ease in a world for which he felt so ill-equipped.

For reasons that will become clear later, neither Judaeo-Christianity nor the dominant secular tradition of post-Renaissance thought had a place for the vitalism of mana, the Tao or the *logos spermatikos* of the Stoics. Over the last century, however, as the traditional structure of Christian belief has collapsed, vitalist philosophies have witnessed a resurgence. The best-known example is perhaps the *élan vital* of

Henri Bergson – popularized by George Bernard Shaw as the 'life force' – but there are many abstruser forms: A.N. Whitehead's 'philosophy of organism', for example. But it was left to ecology – acceptable, as a natural science, to the spirit of Cartesian rationalism – to make it respectable.

In 1942 the American freshwater ecologist Raymond Lindeman proposed a pioneering model for the transfer of energy in a lake. In the Lindeman model light enters an enclosed system, passes through producers, consumers and decomposers, and is emitted as heat. The whole process is represented by an arrowed flow chart.

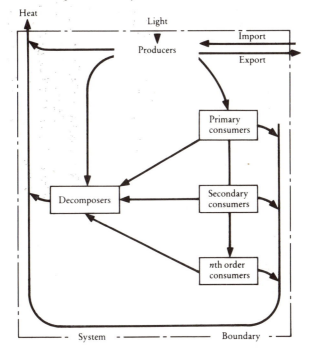

Figure 4.1   Energy transfers in an ecosystem – Lindeman's model

The Lindeman model has been so widely copied as to rank, today, as a culturally generic metaphor for the perception and understanding of living processes. It presaged the way in which an entire community of living creatures, whether it is a lake, a city or the planet earth, can be described, with unrivalled comprehensiveness, in terms of energy flows. Each creature takes up, consumes and releases energy and it is specifically this aspect of its behaviour which relates it to all other creatures, and which integrates and defines the ecosystem of which they are a part.

Yet energy is not a 'thing', a piece of matter like food or a molecule of oxygen, that can be put under a microscope or in a particle accelerator for analysis and categorization. It is a dynamic principle – literally, the capacity to do work – which assumes the form of its host but can assume an entirely different form if its host dies, decays or explodes. In scientific terms it is interconvertible between all forms of matter, its presence denoted by a confusing array of units of measurement. It is also constant: so says the first law of thermodynamics. To sum up, therefore, energy is, in a sense, probably unique to science, immanent, eternal and vital. The implications of this view will be examined later. But two points are worth stressing now. First, the various separate forms of energy were well known by the nineteenth century but it was left to the twentieth to complete the connections between them. The principle of mass-energy equivalence, or mutual convertibility, proposed by Einstein in his famous equation $E = mc^2$ as part of the special theory of relativity, made one vitally important link. Its application to living systems, as exemplified by Lindeman's model and the more recent discipline of community energetics, followed later.

The second point, which bears much repetition, is that for all its scientific currency energy is a metaphor, and a remarkably subtle and sinuous one. The essential meaning of metaphor is that it transcends linear patterns of thought and language to make lateral connections between things which, according to linear rationality, are unconnected: apparently dissimilar things are perceived to be similar, to 'go together'.

As we shall see later, such 'going together' is characteristic of many primitive or magical systems of belief – totemism and astrology are examples. Its perception is of a unity in the universe belying its apparent fragmentation. By convention this is the vision not only of the poet and the mystic but also of the theist; in an existential sense, all things are equal in the sight of God. All things are also 'equal' in the measuring units of energy.

In *The Structure of Scientific Revolutions* Thomas Kuhn argued that profound changes in metaphorical modes of perception – in which, for example, things once thought unconnected are later thought to be connected – are primary agents in the replacement of one system of scientific belief with another. Such changes, a strict interpretation of the scientific method argues, are 'irrational', a product of social and cultural moods and preconceptions. They lead to what Kuhn called a 'paradigm shift' – an upheaval in which one way of visualizing and modelling the world is replaced by another, totally different one. The two, indeed, are 'incommensurable'. The transitions from Aristotelian to Newtonian physics and thence to relativity and quantum theory are two examples. Another, associated with the latter, is the impact of evolutionary theory and ecology on the life sciences.

Compared with its precursors – biology, zoology and botany, for example – ecology integrates, rather than isolates, features of the living world. It deals in cycles rather than notions of linearity. The principle that in the diversity and complexity of an ecosystem lies its stability – that the tropical rain forest, for instance, represents the achieved climax of a long succession, capable of preserving its form in a way that a farm monoculture could not – reinforces aesthetic judgments on the 'richness' of the natural world. It also comes remarkably close to more specifically religious conceptions like that of the perfect plenitude of being radiated by the Godhead. The axiom that particular environments have a 'carrying capacity' relates animals, plants and man to one another, and to survival.

Anthropologists have constantly recorded their surprise at the breadth and exactitude of 'primitive' people's ecological awareness. A dozen neighbouring tribes in the Gabon, for example, were found to possess a botanical list of some 8,000 names. Such systematic cataloguing, according to Claude Lévi-Strauss, meets a need which is far more than practical: it speaks of the perception of correspondences and relationships between human, animal and vegetable worlds, and the spirit(s) that pervades the last two. A Pawnee Indian explains the importance of such ordered knowledge thus: 'We must address with song every object we meet, because Tira'wa [the Supreme spirit] is in all things, everything we come to as we travel can give us help.' Totemism and astrology, similarly, are based on correspondences – 'going together' – between classes of creature, colours, even directions of the compass, which appear unrelated to the outsider. In such cultures the term 'living landscape' assumes a new meaning.

Ecological awareness in the west was enormously accelerated by the 1973–74 oil crisis. Yet the carbon cycle is only one of many cycles that impinge on urban life. Each, in its transition from raw rural resource through consumption to discards and by-products, generates power, prosperity, pollution and waste. Together they constitute much of the environmental agenda, from acid rain to materials recycling. Urban development, particularly in the immediate post-war era, seemed to involve an increasingly elaborate superstructure of civilization capable of 'floating free' of its resource base. Under the impact of cycles-thinking, however, the city assumes common cause with its rural and global hinterland, while the city-dweller, long a pale and spectral creature haunting a succession of interiors – car, home, office, commuter train and shopping complex, each artificially warmed and lit – re-emerges as a peasant, but a clever peasant, his mind brimming with abstractions, his feet anchored in the clay. As Redfield and his colleagues might have expressed it, both city and country express the 'folk culture' of ecology: cities cease to be alien agents of change.

Such cyclical habits of thought, slowly rooting themselves in the western mind, touch directly on the new twentieth-century vision of nature. In 1971, for example, Eugene Odum employed circuit diagrams to express energy flows in human life support systems, depicting the powering of cities by fossil fuels. But this merely echoed the imaginative vision of Aldo Leopold, the American naturalist whose writings have served as a prophetic text of contemporary environmentalism. Land, wrote Leopold, is not merely soil: it is 'a fountain of energy flowing through a circuit of soils, plants and animals'. In Ian McHarg's visionary *Design with Nature*, two decades later, the synthesis between science and imagination, between ecology, evolutionary theory, metaphysics and mysticism, is very nearly complete. McHarg's hypothetical astronaut, originally urban and ignorant, a 'twentieth-century conquistador', has found in his quest for survival that the universe is a 'great creative process' encompassing sun and earth. Its currency is energy, employed in the raising of matter through forms of increasing orders of complexity. Photosynthesis becomes a 'revelatory statement', the world a 'working partnership between the sun and the leaf'. Carbon, meanwhile, is 'the fire at the heart of life'. It is, says McHarg, 'perhaps the most modest creation myth ever advanced'. In 1985, meanwhile, a decade after 'Save It', came 'monergy': the invention of a new word by a Conservative Government, ostensibly as part of an energy-saving campaign, in reality signifying the gradual subversion of an old value system, based on money, by a newer one, based on energy.

Energy, quite clearly, is a reformulation of older, non-Christian forms of vitalism. Like mana or the Tao, it is omnipresent, inherent, life-giving. Hence it is portrayed in terms of heat, light, generation. It represents, like the Stoics' *logos spermatikos*, a principle of order, structuring the universe, yielding an ethical code. Like mana, energy is a metaphor for a process seen as both mysterious and vital: the generation and sustenance of life. It represents also, unmistakably, a version of what Eliade calls 'real existence'.

If the primary response of the individual to the force experienced as wakanda or manito or mana is that it is 'real', however, a secondary response – so cognate, indeed, as to be epistemologically indistinguishable – is that it is sacred or numinous, replete with a thrilling and awesome significance. Such a response is the wellspring of religion. It involves the recognition of a power categorized by Rudolph Otto, in his treatise on the conception of the holy, as *'ganz andere'* – wholly other. Two chief perceptions are defined by Otto: that of the *mysterium tremendum* – the fearful majesty – and of the *mysterium fascinans*. The latter approximates to a sense of the perfect plenitude of being – an idea, as already noted, which also finds expression in concepts such as ecological diversity and the climax ecosystem.

These ideas clarify why the search for a new metaphysic turned to pagan or mystical models. They also shed much light on the developing response to city living.

The starting date usually selected for the decay of modern Christian orthodoxy is 1859, the year of the publication of Darwin's *Origin of Species*. To argue, however, that a nature-based metaphysic, in the guise of ecology, has 'replaced' Christianity would be to oversimplify. Darwinism and its derivatives are merely the most sharply defined of a range of attitudes which have slowly subverted the dominant structures of Judaeo-Christian belief. Feminism, the rise of androgyny and the decay of patriarchalism are prominent among these, and closely associated with the latter has been the undermining of the Christian tradition of transcendence. Similar rhythms in religious behaviour are visible elsewhere in history. The creator beings found in most primitive cultures, for example, ultimately became remote and indifferent and were displaced by gods of energy and fecundity. Examples include Ouranos, castrated by Kronos and then replaced by Zeus, and Jehovah, or Yahweh, frequently rejected by the Hebrews in favour of Ba'al or Astarte. These displacements represent a discovery of the sacred in the tangible powers of existence. Eliade calls this a 'fall into life'.

In the evolution of religious belief animism generally preceded polytheism while monotheism usually came later. There are exceptions, however. An important one is the pre-animistic mana of the Melanesians. This co-existed with a belief in a creator: a being who was, however, so removed from daily life and experience that the islanders scarcely bothered to worship him at all.

Throughout most of western history Christianity, aided and abetted by that other dominant style of metaphysical thought derived from the Greek philosophers, has striven to achieve the very opposite of Melanesian practice. While neo-Platonism inculcated the fundamental and eternal reality of the 'ideal', presenting the material world as a series of mass-produced copies of a single celestial archetype, Christianity simultaneously preached transcendence. The burden of this Christian philosophical tradition was the concentration of 'real existence', of truth and value, into a single unknowable being outside the material world. In contrast to philosophies of immanence, like Taoism or the Melanesians' mana – all of which envisage a *suffusion* of reality throughout the natural world – transcendence robbed that world of meaning, storing it all in the Godhead.

Christianity and neo-Platonism thus conspired together. The result was an endless variation on the ascetics' theme of *contemptus mundi*. At its worst the world was a 'vale of tears', as Browning's *Confessions* expressed it. At its best there remained a residue of poignancy: the appropriate image was the medieval stained-glass window – a brilliant spectacle when illuminated but the light lay elsewhere.

In retrospect, it may seem a Faustian bargain – the surrender of an effectively routine intimacy with nature in exchange for the ontological security and biological uniqueness afforded by the revelation of God in history. For much of the Christian era, perhaps, the price did not seem too high. For medieval man, as R.G. Collingwood has written, even rocks remained animate. The beliefs and practices of the older religions could be incorporated into Christianity, notably at Christmas and Easter, through occultation.

But that there was a price – that a feeling for the beauties of the visible world evoked unease, self-criticism, a sense of betrayal among thinking Christians – is evident in a vast body of writings from the book of Psalms and the confessions of St Augustine to the works of that complex nineteenth-century Jesuit poet Gerard Manley Hopkins. Indeed Hopkins's poem 'That Nature is a Heraclitean Fire and of the comfort of the Resurrection' captures this tension almost to perfection. It is the joy in the world's vibrant energy, in nature's 'million-fuelled bonfire', that initially delights Hopkins. God forms no real part of this but has rather to be invoked *ex machina* to supply explanation and redemption. In purely perceptual terms, in other words, immanence has preceded transcendence.

Hopkins, the first recognizably 'modern' poet, both in his radical experiments with language and in his attitude to wilderness, gives us a unique glimpse of the post-Darwinian mind transfixed by the numinous and attempting to embody it in language. Once Hopkins's world is stripped of its transcendent dimension, however, 'real existence' floods back through nature: a 'fall into life' is produced. The world of hawks, trout and bright winds is, in a literal sense, re-animated.

This, essentially, is the history of Christian religious belief in the century and a quarter since Darwin. Deep ecology, coupled with the advent of systems theory, has transformed the world into a complex network of interconnected life-sustaining cycles, to the point where it can be visualized as one organism. In the Gaia hypothesis, first formulated in 1975 by two scientists, James Lovelock and Sidney Epton, analysis of the history of the earth's atmosphere led to the concept of a 'giant system', regulating temperature, soil acidity and photosynthesis to produce the optimum conditions for its own survival. Such a system could well be categorized a 'living creature' – Gaia is the Greek term for earth goddess – with man as its 'central nervous system', the authors argued.

The cosmos is thus infused with reality. It becomes 'sensitized'. Older attitudes or practices are re-enacted. The Omaha or eastern Canadian Indians, for example, propitiated or apologized to a plant picked for medicinal purposes, since the plant had a soul or was sacred. The prophet Smohalla of the American Indian Umatilla tribe, asked to dig in the earth, replied, 'Am I to take a knife and plunge it into the breast of my mother? It is a sin to wound or cut, to tear or scratch our common mother by working at agriculture.'

In this sensitizing and revitalizing of the cosmos lies perhaps the chief spiritual clue to modern environmentalism. The attribution of feelings to laboratory dogs or rabbits, for example, happens through an act of sympathy – 'going together' in Lévi-Strauss's term – by which humans can in some sense *share* the animals' identity. In Tolkien or Adams, however, the 'going together' has extended beyond the realm of kindred mammals, while anthropomorphism borders on animism: entire landscapes come alive. Hence it is no surprise, elsewhere, to find 'human' rights being advocated not merely for animals but for rocks and trees. In the best-selling *The*

Secret Life of Plants, meanwhile, published in 1974, authors Peter Tompkins and Christopher Bird assert, 'Evidence now supports the vision of the poet and the philosopher that plants are living, breathing, communicating creatures, endowed with personality and the attributes of a "soul"'.

A 'fall into life' of the type outlined represents a major shift of personal and cultural values. It does not, however, imply any straightforward return to paganism: the primitive is recovered and reformulated, not merely regurgitated. Totemistic beliefs may thus reappear as membership of an animal rights group. Tree worship or votive offerings may emerge as purchase of a memorial woodland. The alternative religion of witchcraft, newly resurgent in Britain with between 100,000 and one million devotees, preaches a 'life force' running through everything 'from a red rose to an oil tanker'.

What is common to these responses, both ancient and modern, is the perception of correspondences, relationships, 'going together' – summed up, ultimately, in the vision of a sensitized nature. And what they herald is the almost imperceptible decay and transformation of models of transcendence. Just as a beneficent Christian God slowly replaced the vengeful god of the Hebrews, so that Christian God may ultimately grow into Gaia, assuming so many of the attributes of immanence that he is changed beyond all recognition. The works of figures like Teilhard de Chardin, notably his concepts of the planet's 'mind-layer' and of God as the source of evolution, point in this direction. Or God may, like the Melanesian creator-being, grow increasingly aloof and remote.

The form of the city, as a version of transcendence, and of the unequal distribution of power and meaning that accompanied it, is likely to suffer the same fate. The emerging outlines of its new shape are examined in the second half of the book. Yet as the city's image was projected on to an ever larger screen – including, ultimately, that of the planet itself – old distortions of vision were magnified so that city and countryside became, respectively, emblems of the man-made and the natural. This natural world, flooded with new reality and increasingly charged with the burden of otherness, of the numinous and the sacred, thus becomes a zone of purity and sanctity. It is, in a sense recognized by anthropologists, taboo, instinct with power, offering psychic regeneration to its followers but vulnerable to defilement or 'pollution'. In entering it, one enters a special, different, privileged world, removed from the mundane and the ordinary. In this world one can rediscover and renew oneself. The city, by contrast, was a dead land, a zone of defilement, offering no vision of the numinous and no hope of self-renewal.

Such beliefs are fundamental to environmentalism and help to explain its insistent and sometimes arbitrary distinctions between the natural and the synthetic – in organic gardening, for example. They are increasingly widely diffused throughout western culture, where the synthetic has grown virtually synonymous with the unreal, rather than signifying any tribute to human ingenuity. They are intensely compelling, associated with the deepest sources of religious experience. Yet they also require a space to exist, a place with which in some sense they can be identified.

It is in this context that a geography of the sacred arises, where a psychological need is embodied and expressed in a physical place: an area, in other words, that is both illusion and reality, where the perceptual and the actual merge. It is in this context, also, that the history of the city and that of landscape design assume a vital relation to each other for, as the next chapter argues, the inexorable growth of the city tends to destroy such places and it may be left to human skills in the working of land to recreate them.

(a) Nunhead Cemetery, South London: nature overwhelms art

(b) Planting wild flowers on a reclaimed gap site at Toxteth Liverpool

(c) Dutch experiments with landscape: woodlands surround the post-war tower blocks at Delft

(d) Tree-planting: an act of lifelong significance to a child

(e) An allotments squat on a long-vacant site in Hackney, east London, where prefabs once stood: a street lamp-post (foreground, left) still lies on the pavement

5

A GEOGRAPHY OF
THE SACRED

Could we but behold these things with our eyes as we can picture
them in our minds, no one taking in the whole earth at one view
could doubt divine reason.

*(Balbus, on the richness of creation and the perspective of God, in
Cicero's* De Natura Deorum*)*

'The United States is the most highly advanced scientific and
technological nation and it's paradoxical that, in spite of this, large
segments of the population persist in mythological and primitive
thought.'

*(Paul Kurtz, chairman of the Committee for the Scientific
Investigation of Claims of the Paranormal, on a Gallup poll
showing that 55 per cent of adolescents believed in astrology, 59
per cent in ESP and 69 per cent in angels, November 1984)*

A notable feature of modern anthropomorphism and animism is their alliance with the paranormal – with sorcery, prophecy, strange powers and spectacular beings. Many sources are culled for these creeds and creatures, which when yoked together produce unexpected and often lurid synergies. Tolkien's sentient landscapes, for example, are haunted by elves, orcs and goblins, by ghosts, freaks and half-men, by trees which talk and magicians who make war. They are also inhabited by forces that on occasion seem the purest images of nightmare – forces, in other words, that are straightforward imaginative projections: there is no precedent for them in legend or in the conventions of folk romance. Such worlds are intentionally mysterious – they speak of other dimensions of reality, of other worlds – and self-evidently escapist. Hence they are labelled fantasy. They provide the framework for what is perhaps the twentieth century's most distinctive contribution to the history of literature, science fiction.

They also find expression in a bewildering variety of modern legends which represent, collectively, not merely a publishing and media phenomenon of recent decades but a separate and expanding corpus of pseudo-science. The generic name for this body of work is the 'unexplained'.

The unexplained can take many forms, from spontaneous human combustion to psychokinesis – the ability of mind to move matter without physical connection. But among its most prominent characteristics is the association with unmapped or unfrequented areas of the universe. The Bermuda Triangle, for example, like the legendary lost city of Atlantis, locates mystery – in this case the recurring disappearances of ships and aircraft – in the ocean depths.

The yeti, the bigfoot and their counterparts, colloquialized as the abominable snowman, occupy wild areas of Tibet, China or the Rocky Mountains. The most pervasive legend, that of the unidentified flying object, derives from the enormous blank spaces in cosmological knowledge. It has thus far greater immunity to disproof.

The unexplained displays a hydra-like tenacity. Refute one legend and another springs up in its place. That acute observer of human symbols Carl Jung regarded UFOs as living myth, 'signs in the heavens' appearing in response to pressures caused by lack of living space, the perceived smallness of the planet. Beings from UFOs frequently rescue earthlings – usually from themselves – or point to a better way forward to the future. As in films like *Close Encounters of the Third Kind*, they perform a redemptive function: they are saviour-figures.

Beliefs such as these, now regarded as the preserve of the credulous, were once a fit and fascinating subject of study for intelligent men. Othello, describing his courtship of Desdemona, speaks of his early travels in 'anters [caves] vast and deserts idle', of cannibals, anthropophagi 'and men whose heads grew beneath their shoulders'. The famous *Cosmography* of the German monk Sebastian Münster, published in 1544, echoed the sense of wonder evoked by the voyages of discovery of Columbus, Amerigo Vespucci and others in its portrayals of dog-headed and headless men and one-legged monsters with a single giant foot – creatures supposed to inhabit the New World and Asia. Yet as the long tradition surrounding the so-called *Marvels of the East* demonstrates, this fascination stretched back into classical times. Münster's dog-headed man appears merely a reworking of the *Cynocephalus* – literally, dog-headed ape – of the Romans.

A common feature of these marvels was their residence in *terra incognita* or unknown land – the uncharted and hence 'empty' regions beyond the known world. Among the most representative maps of antiquity was the globe of the Stoic Crates of Mallus, which divided the world into four quarters. The chief of these was *oecumene*, denoting, broadly, the inhabited world and based on the Greek word for home or household, *oikos*. *Oecumene* was centred on the Mediterranean and the Persian Gulf. For the Hellenistic civilization of the west, this was the home of men and the only region on earth which belonged to them, which was familiar and in any sense explored and mapped. The rest was a name, an arbitrary outline and speculation.

The famous *Geography* of Ptolemy, in the second century AD, pushed the mapped world further east, notably to India and the Indian Ocean coastline. Ptolemy's map, however, was only redis-covered in the west in the fifteenth century, shortly before the voyages of discovery. More significantly, medieval maps themselves, mainly produced in monasteries, surrounded *oecumene* with *Oceanus*, beyond which frequently lay another region of earth where monsters and fabulous beings lived. Here, also, Paradise was located. Such trans-Oceanic vastnesses were commonly designated *terra inhabitabilis* and denounced as perverse and sinful. They thus bore a close resemblance to older conceptions of wilderness, a term most commonly linked in Norse and Teutonic usage with the forests that covered northern Europe in the wake of the withdrawing ice-sheets.

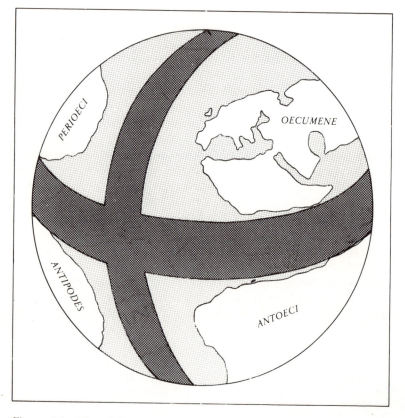

Figure 5.1   The globe of Crates of Mallus

The forests, as Roderick Nash points out, were in the earliest times thought to be the home of supernatural beings: satyrs, centaurs, fauns, trolls and werewolves. But their most notable resident was the wild man – a semi-human being of great strength and sexual potency found in one form or another throughout European legend and folklore.

Terra incognita, however, was inhabited not merely by the supernatural, the fantastic and the profane. The medieval monks' siting of Paradise beyond *Oceanus* was accepted even by St Thomas Aquinas, greatest of the scholastic philosophers. Paradise, said Aquinas in his *Summa Theologica*, lay in the east. So far it was undiscovered because it was shut off by mountains, seas or impassably torrid zones. So pervasive were such beliefs that Columbus died believing he had found Paradise on his third voyage. Haiti, he was convinced, was the Biblical Ophir.

The creations of men's faith as well as their fears and imaginings were thus seen as having a physical dwelling-place on earth – a conception that extended to God himself. As E.M.W. Tillyard wrote, God was imagined by the ordinary Elizabethan to be 'domiciled beyond the bounds of the fixed stars . . . attended by the hosts of the angels'. This belief was intrinsic to the great chain of being, a picture of the world with its origins in Plato's *Timaeus* which was a commonplace of thought until the eighteenth century.

Some important conclusions can be drawn from men's map-making activities. First, the sacred and the fantastic have for most of history evidently shared the same space in the human mind, a space represented on the map by *terra incognita*, the world 'out there', elsewhere, the 'other' world beyond that of home and habitation (*oecumene*). The modern disciplines of phenomenology and cognitive psychology have helped us to understand the crucial importance in the development of knowledge of the interaction between an organism and its environment. They have also given us new terms to describe it. Chief among these is the 'mental map' – an image or structure which exists in the mind but which both replicates and shapes the world as experienced. Maps, from the three-quarters-empty globe of Crates of Mallus to the immensely detailed products of the Ordnance Survey, thus describe the mind that made them as much as the terrain they chart. They represent a modelling of experience into a physical form in a similar manner to a painting or a sculpture.

But since minds make maps and maps make reality, maps can

introduce new orders of reality in response to mental constructs and configurations which can nowhere be seen or sensed – which do not, in other words, belong to the 'real' world. The shape of the five continents was known to people long before they were first seen and photographed whole by the Apollo astronauts in 1968. Even now the detailed geophysical, political and economic picture provided of the earth's surface in atlases is an illusion, manufactured by technology for the purpose of a highly complex and specialized culture. And since these pictures are our only representation of the particular version of reality we have selected as paramount, they *are* that reality. Asked to describe a country or a continent, a westerner would describe a map.

Different minds and cultures, however, produce different maps. The Aleutian islanders of the Bering Sea have names for the tiniest of streams or ponds but none for their mountain ranges. Nor is there a generic native name for the islands themselves since they 'do not recognize what to us appears to be the obvious unity of the chain'. Young children, as developmental psychologists have shown, exist in a universe so proximate, perspectiveless and pre-abstract that to turn a corner out of their sight is to vanish and to reappear is miraculous. Interestingly, we praise them for their imaginative qualities without realizing that it may well be in those empty and unknown spaces round the corner and beyond the closed door that their imagination lies.

Wilderness and *terra incognita* thus occupy both mental and physical space: the two categories, in an important sense, are indistinguishable. For much of history, the mind's wilderness and the world's wilderness have acted as mirror images of each other. Inner and outer landscapes were in harmony. In both dwelt the marvellous, the monstrous and the sacred. Looking out from the boundaries of home and inhabited world, western man saw stretched before him a vast realm of infinite mystery and potential. The experience must have been peculiarly and paradoxically satisfying, giving a sharper sense of what it meant to be human, a greater joy in men's own works and company, a greater fear and awe of what lay beyond. The disjunction of subject and object which is so intrinsic an element of contemporary fantasy simply did not arise on anything like its modern scale.

The great illusion of maps, moreover – that they represent an objectively real world – meant that this realm of mystery, of the 'wholly other', was available to people through the simple medium of travel. In other words, while *terra incognita* could not necessarily be reached by means of a simple road or sea trip, the journey was theoretically possible: only geographical obstacles lay in the way. And as late as the eighteenth century Pope can be found expressing in his *Essay on Criticism* the still widespread belief that the residences of the angels would in time be located, possibly as new planets were discovered.

Several events conspired to make such imaginable physical contacts increasingly unlikely. The forests, covering over two-thirds of Britain's total land area in Neolithic times and a third in the post-Roman era, shrank inexorably, covering less than 5 per cent at the start of the twentieth century. The voyages of discovery had the effect, ultimately, of drastically reversing the balance between *oecumene* and *terra incognita*: the former came to preponderate over the latter and, most significantly, to *surround* it. Advances in astronomy, coupled with the Copernican substitution of a sun-centred for an earth-centred universe, led to the unpicking of the great chain of being, so that God could no longer have his domicile beyond the fixed stars in the outermost of the eleven spheres that ringed the earth.

Although this was undoubtedly a slow and cumulative process, it is in the later eighteenth century that the notes of unease begin to cohere into a recognizable chorus of dissent. Civilization – that is, the inhabited world, or *oecumene* – becomes a term capable of expressing an ill. Cities, the distilled essence of civilization, start to proliferate and concentrate. The landscape gardeners attempt to call in the country, inventing the ha-ha to make the distant horizons, with their hint of unexplored emptiness, psychologically available to the spectator. And the garden, the very type of magical and privileged enclosure, loses its status as special space, set apart. It becomes no longer a presence, more a representation of nature.

It was in this context that the two dominant traditions of western culture, Judaeo-Christianity and scientific positivism, conspired to rob the proximate world of significance. Christian transcendence taught that true otherness lay elsewhere, not in the here-and-now. Science, meanwhile, insisted that if otherness did not exist in physical space, particularly in the physical space of *terra incognita*, it did not exist at all.

The consequence was a profound change in attitudes, associated particularly with the notion of boundaries and limits and exemplified by the altered status of the walled garden. The original sense of a garden is precisely that of enclosure, a term which describes some of the most set apart and hence privileged places in legend – not only Eden but the more generic Paradise, itself derived from *Pairidaeza*, meaning enclosure or park in Avestic, the ancient Persian language of Zoroastrianism. In the walled garden of medieval poetry, courtly romance blossomed: the restraints and concerns of the real world were excluded. Yet by the early eighteenth century it was the walled formality of the contemporary Dutch garden, its sense of restriction and constraint, that led to the reaction among British designers which gave birth to the landscape movement.

The point becomes clearer by examining the design of a traditional Japanese garden, imbued with the philosophy of Buddhism. This was intended to be a small landscape in its own right, not an imitation of a larger landscape elsewhere. Often it is surprisingly formal. The gardener trims, prunes and weeds 'out of a feeling that he himself is part of the garden . . .. He does not have a disturbing influence on nature because he *is* nature.' The garden, in other words, is both highly artificial *and* highly natural. A crucial boundary thus appears to vanish.

The global balance between *oecumene* and *terra incognita* was reflected more intimately in that between house and garden. The relationship, however, is a complex one: inspecting it is rather like unpeeling an onion, each layer involving a different psychological response. Surrounded by wilderness, antiquity sought a nature in which human design was apparent, since this was clearly not the case in the world beyond. Within men's own settlements the imbalance and incomprehension of this wider world could be redressed: home could embrace wilderness.

Hence there evolved not only the public park of Nineveh, Babylon, Hellas and Rome – parks which would now be viewed as highly ornamental – but the many forms of small enclosed garden which perfected the harmony of the architectural and the natural. Beginning with the peristyle and the atrium – Greek and Roman variants of the inner courtyard – the tradition continued with the medieval monastic cloister and the water-haunted 'paradise' of Byzantine and Moorish palaces. All were outdoor rooms, delicately integrated into the physical fabric of the home.

The achievement was that of a seasoned urbanity, long experienced in communal living. When this perished, balance and integration were also lost. And when a new urbanity was needed to cope with cities of unprecedented size, these two key elements of it, house and garden, had become separate and autonomous realms. Compromises – verandahs, conservatories, gazebos – were devised but by the twentieth century the apartheid had grown so rigorous that high-rise flats dispensed with gardens altogether. The delight in the near-at-hand, the fear and awe of the world beyond, had both been comprehensively eliminated.

The final note in that eighteenth-century chorus is a peculiarly familiar one: the notion of global limits which so powerfully impels environmentalism. It is struck most notably in a tract often cited as an early classic of ecology and which had an enormous influence, Thomas Malthus's *An Essay on Population*, published in 1798.

Malthus's central argument is that the existence of unpeopled lands merely postpones the inevitable conflict centred on a growing population and a static supply of land and food. By any standards, it is a remarkable anticipation of the debates of today. The image upon which it relies, however, is even more striking. 'A man who is locked up in a room,' said Malthus, 'may be fairly said to be confused by the walls of it, *though he may never touch them . . .*' (author's italics). Not only is the earth's finitude here portrayed in oppressively claustro-phobic terms, a notable early example of such a decisively modern usage, but the image precedes and shapes reality. The interior of Africa remained to be explored, as did the two polar continents. Japan and China were still largely unknown to the west. Cognitively, however, Malthus had leapt forward in time, discounting future discoveries as a form of mopping-up operation. *Oecumene* had finally surrounded and circumscribed *terra incognita*.

The image of a mapped and bounded earth, powerfully enhanced by the blue and white orb which appeared in countless newspapers, magazine and television pictures as a result of the Apollo space programme, is now taken for granted. It is part of the mental equipment of every educated westerner, along with terms like 'spaceship' or 'planet earth' and the idea of the 'global commons'. Map-making has become a craft of almost pinpoint precision. Satellite technology, for example, can now measure continental drift of one or two centimetres a year.

Yet the mind that requires such detailed knowledge also revolts against it. Jung saw UFOs, for example, as a sign of humanity wishing to 'escape from its prison', an image remarkably similar to Malthus's. And the United States at the turn of the century produced a decisive practical demonstration of the psychological factors involved. The 1890s was by convention the decade in which the

American frontier, and with it an era of discovery and revelation, ceased to exist. The sense of fatefulness and loss was national. The very existence of the frontier, wrote its historian Frederick Jackson Turner, 'appealed to men as a fair blank page in which to write a new chapter in the story of man's struggle for a higher type of society'. Aldo Leopold expressed it more pungently. 'Of what avail,' he asked, 'are 40 freedoms without a blank spot on the map?' There is in retrospect an exceptional irony in this spectacle of a supremely positivist culture transfixed and helpless in the face of its own unyielding image of the world, an image in which exploration was always outwards and elsewhere, towards boundaries that refused to move.

The retreat of the mysterious and the marvellous under the advance of cartographic man has taken many forms. God, for example, has become variously mysticized or etherealized. His traditional home in the heavens, the empyrean of Milton's *Paradise Lost*, has receded to an unimaginably remote and impossibly speculative point before the big bang and the onset of Einsteinian space-time. The cynocephalus and its relations have made their newest home in Hollywood, where they feature in multi-million dollar epics of the occult, the monstrous and the extra-terrestrial. As physical discovery gives way to mental escapism, *terra incognita* has grown more eclectic and spectacular, sprouting a profusion of bizarre and increasingly private fantasies. Like all formerly unspoilt regions, it is heavily commercially exploited.

The relevance of myth, marvel and mystery to psychological health was powerfully argued by Jung and is now implicit in much psychotherapy. In Jung's theory of the collective unconscious – a buried psychic world common to men and thronging with the creatures of their imagination, embodied in archetypes – *terra incognita* is fully internalized. The collective unconscious is the mind's wilderness. Jung described it as 'the hidden treasure upon which mankind ever and anon has drawn, and from which it has raised up its gods and demons, and all those potent and mighty thoughts without which man ceases to be man'. It is the source of myth, of religion, poetry, folklore and fairy-tale. When man loses touch with it, 'he loses touch with the creative forces of his being'.

The retreat into inner space, however, carries a special penalty. Jung himself spoke of the danger of 'succumbing to the fascinating influence of the archetypes' (of the collective unconscious). At some point the world of objective space, the space of homes, travel, leisure and work, has to be re-entered and re-inhabited. As the two worlds become increasingly difficult to reconcile with each other, the psychological and cultural dislocation intensifies. For Jung, the chief dislocation lay between the conscious and the unconscious. Personal wholeness, or individuation, lay in an integration of the two, a theme common to modern psychotherapy. Similarly, city and country, settlement and wilderness, *oecumene* and *terra incognita*, represent aspects of the human mind which have come to seem irreconcilable opposites, a dualism immobilized in living patterns, in the distribution and density of population and industry. And since houses, roads and workplaces are built to last, men can fairly be said to inhabit the minds of their ancestors.

The landscape of the twentieth century was in large measure the invention of the nineteenth, which is to say it is beset by a sense of dislocation which borders on the compulsive and the neurotic. If there is such a thing as a typical Victorian reaction to the city, it is not so much that of Dickens, who used the city as a metaphorical device to explore an entire society, as of Richard Jefferies, the Wiltshire farmer's son turned journalist and nature writer, and it is seen as its most characteristic in Jefferies's novel of 1885, *After London*.

After London is a compelling example of what the American psychologist Martha Wolfenstein has called the 'post-disaster Utopia' syndrome, in which apocalypse – the collapse of a tainted world – is followed by an era of new hope. In Jefferies's novel, however, the disaster, although prophetically environmental, leads to an oddly regressive medievalist fantasy. London's end comes by drowning, under a spreading inland lake. Its corpse remains, however, a rotting and acrid marsh exuding foul taints and vapours. These assume ghostly human shape and yet retain a virulent and indestructible toxicity. Jefferies's London, indeed, resembles nothing so much as a medieval vision of hell – the works of Bosch and Dürer spring to mind – in which modern chemicals have become the means of torture and corruption. From it his knightly hero flees, finds shelter and sleeps, waking to find himself among scuttling moorhens and a warbling thrush on a tiny green island far from the marsh. 'It was like awaking in Paradise,' comments the narrator.

By contrast the characteristic twentieth-century reaction is less anguished, less sharply focused. Its target, from the foundation of the Council for the Preservation, later Protection, of Rural England in 1926, through the strictures of Nairn and like-minded conservationists, to the imaginative worlds captured by Tolkien and Richard Adams, is the spread of urbanization. Ultimately, although its anger scarcely abates, there is an ominous sound of resignation and hopelessness. Each reaction carries its particular truth. The typical Victorian city was densely packed but relatively small. The mid-twentieth-century city was larger but more diffuse. More particularly, the spread of the latter was becoming a metaphor for the colonization and crowding of the earth itself. *Oecumene* had developed into *ecumenopolis*. Spaceship earth was now a city. The planet itself was settled and civilized, threatened with the status of artefact. An entire continent was a 'world park' – as Greenpeace, following the government of New Zealand, designated Antarctica in 1985.

The result was that although the crowded urban millions of the nineteenth century were denied escape as never before, escape, when it came, seemed to be into another realm of being – the 'paradise' of Jefferies. For the twentieth century, nagged by thoughts of a spreading city and a charted planet, escape became increasingly difficult. Settlement was present even when it was not visible. It was around the corner, over the hill, beyond the horizon. Everything was slowly *spoiling*, a theme explored most single-mindedly and popularly in the poetry of John Betjeman but perhaps most tellingly in Philip Larkin's poem 'Going, Going':

> Despite all the land left free.
> For the first time I feel somehow
> That it isn't going to last
> ... And that will be England gone.

Larkin's poem, notably that despairing word 'somehow', captures exactly the sense of mystified unease, the inability to pinpoint a precise cause. Hence the observed paradox of megalopolis – the spectral presence of a city where, physically, no city exists. But the city did exist, in the mind.

A challenge presented itself, one which took many forms. In part it was that of unmaking megalopolis, of reconstructing the city of the mind in such a way that it ceased to exist or, if it continued to exist, it ceased to persecute. Central to this was the old division between *oecumene* and *terra incognita*, and the new, immanentist view of nature. If society retained the distinction between the settled and the wild in its historic form, the wild would simply vanish, and with it, if Jung is to be believed, the source of men's creativity and psychic energy. Moreover, if only what was 'natural' was ultimately real, reality would be banished along with the wild. A desperate search for spirituality would be launched, condemned at its beginning to failure, since what was sought was, by definition, no longer available. The implications were considerable. 'All things,' declared Sir Thomas Browne, 'are artificial; for Nature is the Art of God.' Christianity, for all its denial of the proximate world, offered a vision of ultimate reality, of God, made manifest through God's own 'artificial' works. Equally important, Christianity offered the immensely comforting prospect of power over nature without responsibility. However man might treat nature, it remained not his work but God's. It was God's image he saw in it, not his own reflection. With the passing of God, that comfort has gone. Man stands alone, confronting the planet, able to remake it in his own image but deeply fearful of the consequences – the chief one that of never being able to escape from himself.

The use of land and living-space provided the key since it was here, in the 'real' world, that men might create the refuges their minds no longer allowed them or might discover that they no longer needed them. The 'fall into life' signified by immanence also indicated a new companionability between man, animals, plants and the natural world which might, in turn, point to a more equal cohabitation of the planet: a partnership with nature. The landscapes of legend and literature, of Tolkien and A.A. Milne and Kenneth Grahame, might become places to live in. The 'man-made wild' – the term coined in 1970 by Nan Fairbrother – might become reality.

Since a man-made wild is in a strict sense an impossibility, the obstacles to be overcome were clearly substantial. Jung appeared to believe it could not be done. Echoing Otto, he remarked that the psychological structure of the religious experience was confrontation with the 'psychically overwhelming other'. The redemption of the individual in a mass age could be achieved by 'numinous transformation ... a kind of metaphysical command'. He was quite clear this could not be 'manufactured artificially'.

The pages that follow attempt to show how people have responded to these challenges and thus whether Jung was right.

THE PARABLE OF THE BOG

The country in which I lived in childhood was being shabbily destroyed before I was ten, in days when motor cars were rare objects . . . and men were still building suburban railways. Recently I saw in a paper a picture of the last decrepitude of the once thriving corn-mill beside its pool that long ago seemed to me so important.

(J.R.R. Tolkien, Foreword to The Lord of the Rings*)*

Real life has an uncanny knack of imitating fiction. In *The Lord of the Rings*, the company of hobbits, having left their homely Shire and entered the Old Forest, find themselves led against their will into the deeper parts of the forest by a landscape that seems possessed of sentience and purpose. They end in a dim-lit gully overarched by trees, then abruptly stumble through a cleft in a high bank upon a valley bathed in golden afternoon sunshine – a hidden land, warm and drowsy. 'In the midst of it there wound lazily a dark river of brown water, bordered with ancient willows, arched over with willows, blocked with fallen willows, and flecked with thousands of faded willows.' It is the valley of the River Withywindle, the centre and source of the wood's 'queerness', of the old-grey willowmen who envelop and imprison the hobbits, but at least a temporary refuge, a place of reassurance in a threatening world.

Or is it? Is it, instead, an eight-acre fragment of marshland three miles from the heart of modern Birmingham, hidden behind mock-Tudor semi-detached houses, once part of an old heath called Greet Common, later – at the beginning of this century – named Happy Valley, also christened the Dingle and now known, more precisely but less romantically, as Moseley Bog?

Moseley Bog is a site of special scientific interest, a nature study area and the focus of a remarkable spontaneous protest which developed into organized community action. It is also a parable.

The history of Moseley Bog is a fascinating mosaic of old habitations, industrial, human and ecological. The heath, marsh and meadows of which it once formed a part, marked by local place names like King's Heath, Wake Green, Swanshurst and Coldbath, fed water-mills and fish ponds in medieval times. Old Pool, now the site of the bog, was made as a reserve to feed Sarehole Mill. The land was

Plate 6.1   Moseley Bog: secrecy and magic

not fully taken into private ownership until Victorian times, when the spread of Birmingham out into the countryside of Warwickshire and Worcestershire began in earnest. First villas were built, their sprawling, exotic gardens stretching down into the valley, and then this still recognizably rural outpost of the city was enveloped in twentieth-century housing. Old Pool was neglected, leaking out and waterlogging the surviving meadows around it, and in 1919 Sarehole Mill finally closed down.

In the late 1890s, a decade or two before the valley was swallowed by the city, a 5-year-old boy came to live at 264 Wake Green Road with his sister and recently widowed mother. Beyond their back garden fence lay the valley, an unremarkable little landscape of woodland, wild flowers, small streams and damp hollows, and further down Coldbath Brook, along Old Pool Meadow, was the mill. The two children, naturally enough, climbed the fence, explored the woods and streams and even gave the miller's son at Sarehole a nickname – the White Ogre.

The boy was John Tolkien. The family spent only four years at Wake Green Road but they were, he explained many decades later to his biographer, 'the longest-seeming and most formative part of my life'. Tolkien the author stoutly resisted attempts to read political or allegorical significance into *The Lord of the Rings* but he did concede that it had 'some basis' in experience. The experience was that of 'shabby' urbanization – and with it the retreat of some charged and magical realm in which a lost countryside had become a lost childhood, and a decrepit corn-mill could stand as symbol for both.

To repeat the question: is the valley of the River Withywindle, or indeed Lothlorien or Rivendell, or any other of the enchanted places abounding in *The Lord of the Rings*, merely, or really, Moseley Bog? The answer, of course, is: yes and no. The Withywindle valley is clearly far more than the wood at the bottom of the Tolkiens' garden. It is that wood as Tolkien the writer, recalling Tolkien the child, remembers it or would like to remember it: a wood of magic and mystery, a wood, in Otto's word, of 'otherness', where the watcher is awed and entranced by the sense of a power beyond his imagining. The forms of a landscape thus assume a symbolic quality, become the embodiment and expression of a human need as well as objects compounded of water, air, rock and carbon.

Part of the parable of Moseley Bog is, indeed, one of the lessons of landscape history and cognitive psychology: that the observer sees not what is really 'there', but what he has been taught to see and wants to see. To paraphrase Kant, since reality is mediated through men, one cannot separate the process of knowing from that which is known: 'things in themselves' – *noumena* – are thus unknowable.

Art, literature and landscape, in fact, are hopelessly intertwined, which explains why today, despite the passage of two millennia, we 'see' the world around us with the eyes of the classical Latin poets. Modern landscape painting originates primarily in Holland and in Italy, and notably in those sun-lit arcadian landscapes of the Roman Campagna painted in the seventeenth century by Claude and Poussin. These were widely assumed by an age well versed in the classics to be illustrations of the idyllic pastoral life described, for instance, in Virgil's *Georgics*. Richard Wilson, the first in a line of great British landscapists who still supply some of our most potent images of the countryside, was sufficiently inspired by these visions of the Campagna to abandon portraiture for landscape painting in 1750. For Pope, meanwhile, 'calling in the country' meant, precisely, laying out physical landscapes which replicated Claude's pictorial landscapes.

This object appealed perfectly to larger and smaller aristocrats, lesser gentry and greater suburbanites, who fancied themselves as the *beatus ille* of Virgil or Horace – the happy man living in gentlemanly rural seclusion.

Variations on these eighteenth-century landscapes, in the shape of stately home, public park and private garden, now surround us, while a recognizable Virgilian Arcadia smiles at us from that most reproduced of paintings, Constable's *The Haywain*. Travellers of Constable's period, of course, surveyed the view with their backs turned on it, gazing instead into a plano-convex mirror known as a Claude glass which framed the scene as tidily as in a painting. The history of man's relationship with his surroundings seems at times an affair of borrowed perceptions refracted through artificial media in which things in themselves – what we like to think of as the real world – scarcely intervene at all. In landscaping, even more than in architecture, art merges with life. It thus becomes inescapable.

Just as art imitated life in Tolkien's transfiguration of Moseley Bog into the Withywindle valley, life now returned the compliment. In the late spring of 1980, a party of schoolboys and staff from King Edward's Grammar at Aston visited the bog to continue their work of restoring it and surveying its wildlife. By now the Victorian villas had vanished and their garden exotica, mingling with the original wetland plants, had produced an idiosyncratic flora in which bittersweet, skullcap and bog pimpernel were found alongside Japanese maple and flag iris. The bog was home to two relative rarities, royal fern and the reed-like wood horsetail. To most local people who wandered through it in their wellingtons with the dog, however, it was merely a damp patch of forgotten woodland, full of mossy logs and fungi and woodpeckers and blackberries for picking in the autumn. But on that visit in 1980 the school party found surveyors tramping through the old gardens leading down to the valley, sizing up the terrain for the construction of twenty-two detached houses.

If the incident recalls in essence and also in many of its particulars – the time of year, for example – the opening of Richard Adams's *Watership Down*, or the 'scouring of the Shire' at the end of Tolkien's epic, this was much more than coincidence. That distinctive twentieth-century experience of apparently inexorable urbanization was repeating itself. But the events that followed showed what lessons life had learnt from art.

A campaign of resistance immediately sprang up. Its chief begetter was Mrs Joy Fifer, a housewife and mother of four grown-up children who, like Tolkien, lived a few hundred yards away on Wake Green Road. It began like a whirlwind and ended in a steady breeze. Within three weeks in June 1980, 12,500 signatures were collected, 500 protest letters were sent to the council and 300 people had turned up at a public meeting at a local school hall. 'I'd never attended a public meeting before, let alone chaired one,' Joy recalled much later. 'I was so scared I was actually shaking.'

The Save Our Bog campaign then moved into full swing. Hobbit balls and a *Lord of the Rings* read-in were canvassed, a teddybears' picnic organized, and an impromptu choir formed of local children who played around the bog. Gradually the protest assumed an air of solidity, as of new forces taking shape. The campaigners' bulletin, *Moseley Bog Paper*, was by the summer of 1980 already planning a survey of wildlife in Moseley. And it finished with a footnote on the subject of dreams.

Over the first few weeks, it said, the people who wanted merely to save the bog had learnt a lot from the professionals they had consulted. What about getting the 'experts' and the 'people' together during the long winter evenings to come? 'We could begin to learn how Moseley Bog works, from the geology through the flora and fauna to people and how they have used it, in the past and now.'

The battle of Moseley Bog was not won finally, and then not totally, until three years later when the Roman Catholic Diocesan Schools' Commission, which wanted to develop the land, was forced to accept a drastically reduced plan involving only eleven houses, leaving the precious wetland largely intact. By then, however, events in Birmingham, and indeed in many other cities, had moved on: what might have been an isolated protest of slightly above-average duration had taken on a much wider significance.

For the next lesson in the parable of Moseley Bog is the sense of wonder that it was there at all. 'I've lived only a mile away from it for eight years and I've only just learned of its existence,' wrote one local journalist. Underlying the enthralment of the many visitors – journalists, councillors, conservationists, residents – who came to see it was one simple theme. What they *said* was that Moseley Bog was an 'oasis of beauty', a 'precious vestige' of a vanished West Midlands, a forgotten patch of nature which had 'miraculously escaped the ravages of progress'. But what they *meant* was that Moseley Bog had *no right to be there.*

This was an extraordinary and intriguing response. The reasons for it are highly instructive.

The history of contemporary Britain is usually taken as dating from the last war and the years immediately following it, when a new welfare state, education system and public sector bureaucracy was created and Keynesian principles of a managed economy accepted as the prevailing orthodoxy. A similar consensus was elaborated over the treatment of city, country, land and planning. Both settlements, significantly, have since the 1970s been called increasingly into question.

The land-use settlement was formulated in a series of reports and new legislation during the 1940s but was in important respects a culmination of trends dating from the industrial revolution. It was also a response to wartime food problems and to the rampant urban sprawl of the 1920s and 1930s. In brief, it gave us national parks, new towns, the green belt, a regional jobs and population policy and a comprehensive planning system. Food production was to be maximized and agricultural land, left outside planning controls, treated as sacrosanct. Sir Patrick Abercrombie's wartime plan for London, envisaging a green belt ten miles across, served as the model for much city planning to follow.

The beliefs of some of the leading figures involved in this settlement are illuminating. Sir Dudley Stamp, for example, probably the best-known geographer of his generation and founder of the first Land Utilization Survey of Britain, viewed the farmer as the nation's unpaid landscape architect. Patrick Abercrombie, that most influential of modern planners, thought that 'the town should ... be frankly artificial, urban; the country natural, rural'. And the men whose preferences lay behind the choice of mountain, moor and clifftop for the ten new national parks created in 1949 – writers like Vaughan Cornish, hill-walkers like John Dower – were impelled by a vision of solitude, of man communing alone with nature in an otherwise empty landscape and drawing strength from such communion.

That, indeed, appears to be the essence of the wilderness experience – not so much an aesthetic response to the beauty of landforms as a social and psychological response to the absence of other people and their works. A 1981 survey of American wilderness-users – backpackers, canoeists and so on – found, for example, the following qualities listed in descending order of importance: tranquillity, absence of man-made noise, freedom of action, 'being yourself', freedom from man-made intrusions, 'freedom to control your thoughts' and, lastly, a 'completely natural environment'. But perhaps the most intriguing feature was the failure to mention the actual physical texture of the landscape in which these solitary backpackers invested such lengthy reveries.

Particularly over the last three decades, cognitive psychologists have shown how people not only see the world through mental maps but also divide it into apparently distinct psychological realms. In a study of Devon primary school children the British psychologist Terence Lee found evidence of two mental models, or schemata. The first was the home area – various local schemata which bore a 'detectable relationship to the physical world'. The second lay far beyond – one total schema in which physical dimensions were irrelevant and which Lee described as the 'elsewhere' schema. Another study on the way in which children learn to orient themselves in an alien world by constant reference to their home – so-called 'domicentricity' – found that their trips to faraway places were remembered 'but not thought of as connected with, or of the same world as, their immediate habitat'.

In the elsewhere schema can be recognized an authentic descendant of *terra incognita*, a similarly potent source of fantasy and similarly regarded as the antithesis of 'home' (*oecumene*). Both act as a kind of blank screen upon which the mind can project some of its deepest needs without interference from recognizable reality. And the post-war land-use settlement redrew the landscape of an entire nation in their image. Home – which, since 1901, had meant cities or towns for at least four-fifths of the population – was largely discredited. Only elsewhere, in a world increasingly unrelated to the everyday, was meaning to be found. Cities, viewed as unhealthy, destructive, worn and tarnished, were not only to be shut in upon themselves by the green belt but were to be subjected to the most detailed scrutiny and regulation by the planners. Farmland, by contrast, was 'open background' and was thus left as a blank space on the development plans. It had thus been collectively designated, like the national parks, an 'elsewhere' schema.

Moreover, the new land-use patterns and resulting lifestyles tended to exaggerate the caricatures of city and country which had produced them. At least between the wars, as charted by Dennis Hardy and Colin Ward in *Arcadia for All*, ordinary people with very little money could find plots and build houses for themselves by the sea or in the country. Planning controls not only outlawed the plotlands: they made a place in the country prohibitively expensive. In the 1930s the plotlanders could buy their 20 feet by 100 feet lot for £109 an acre. By 1984 building land in the London green belt was slightly over £260,000 an acre while housing land sales in Berkshire during 1985 produced prices of £350,000 an acre. In the half century to 1985, comparing the plotlanders' outgoings with the prices touched in Berkshire, the cost of land for housing rose 137 times faster than retail prices. In the single decade 1960–70, meanwhile, land in outer London increased as a proportion of the total house price from 10 per cent to almost 40 per cent in places – a level that by 1985 had become common throughout the South-East.

The chief reason for this was a market shortage of land, caused not only by zoning and development control but by the theoretical inviolability of green belt and farmland. One result was to drive city-dwellers into the hands of the local council and thus up into tower blocks, the easiest solution to soaring urban land prices when combined with fixed standards and cost ceilings for municipal housing. Another was that British people generally, as the post-war years wore on, found themselves living in smaller houses with smaller gardens. By the mid-1970s a discovery was made, and a new label invented, by researchers surveying, for the first time in the post-war years on such a concerted scale, the inner areas of cities. Lambeth in London was one example. Many people were 'trapped' there, deeply dissatisfied with city life, dependent on welfare yet too poor to move away, without a car and often without the money even for a day trip. A survey by the Countryside Commission in 1973 at country parks and beauty spots just outside Greater London – places like the Lullingstone Roman villa in north Kent, fifteen miles from the city centre – found no visitors from the deprived inner areas. Most Londoners there were from the nearby outer boroughs.

Whether or not it is entirely fair to call these trapped inner-city-dwellers prisoners of the planning system, they do have a story to tell, and it is one which involves millions of people, perhaps the better part of a nation, embarking on a prolonged tail-chasing exercise. The cue for this was the doubling of car ownership in the 1950s and 1960s and its consequence was panic amongst the planners. Michael Dower's booklet of 1965, *The Fourth Wave*, summed up their mood, with its evocation of scurrying ant-like millions 'swarming out of cities in July and August', threatening coast and country alike. This was the era of the day tripper – an altogether different creature from the rosy-cheeked and khaki-shorted rambler and cyclist of the inter-war years – and although the planners' worries were overdone, they culminated in the setting up of the Countryside Commission in 1968 and the creation of scores of country parks to act as 'honeypots'.

As well as a Beatles song, that era also has an eloquent literary epitaph in the shape of Nan Fairbrother's award-winning *New Lives, New Landscapes*, published in 1970. The Fairbrother thesis was that nostalgia should not be allowed to interfere with the landscapes created by new technologies and lifestyles. Among its most potent images was that of a middle-aged couple evidently newly arrived on just such a day trip as *The Fourth Wave* envisaged. The photograph shows them sitting in a wood, on folding chairs, drinking tea from a flask, the man in jacket, tie, pullover, trousers, the woman in neat, woollen suit. Through the wood can be glimpsed, vaguely, the outline of car and road. 'No need of boots or binoculars, to appreciate the country', the caption reads, but the true story, belying such breezy optimism, is quite different. It is that of strangers in an unexplored land, convinced that there must be some qualities of interest or novelty in it but completely unable to locate them. An atmosphere of wariness is also present, so that although the wood is in Swaledale, Yorkshire, it might almost as well be the far side of Pluto, with the couple's car a form of interplanetary life capsule ready for instant take-off should danger threaten.

New Lives, New Landscapes, as a book with much to say about the future of land-use and landscape design, was acclaimed on publication. Yet so much did attitudes change in the 1970s that within ten years it seemed seriously dated. Miss Fairbrother's middle-aged couple find themselves psychologically stranded in Swaledale largely because of what she describes as a policy of 'essential importance to landscape' – keeping town and country as separate environments. The car is their chosen method of connection between the two and it is also Miss Fairbrother's. Roads, she says, can be beautiful. Britain is highly industrialized. An urban population with high living standards demands motor cars. 'A negative policy of not disturbing the old cannot . . . for long succeed. We must disturb it to survive – on a vast scale and everywhere.'

There is much here that is true yet the underlying vision remains flawed. Miss Fairbrother's car borne pursuers of emptiness and glum starers at 'fine scenery' – which, she adds unhelpfully, is 'precious because it is beautiful' – are trapped by the world of 'elsewhere'. Her Swaledale couple are evidently not wilderness enthusiasts of the backpacking variety. But they would certainly have found within themselves a deep cultural resonance to the marketing imagery employed by the British Trust for Conservation Volunteers.

The BTCV, founded at the height of the cold war between town and country in 1959, is one of the most effective of those traditional voluntary organizations which serve as both a symptom and a putative cure of the divide: a sort of psycho-social sticking plaster. This is because they offer townspeople an active function in the countryside – from scrub clearance to dry-stone walling – beyond that of mere inspection. They provide, in other words, a stronger connection. Their rhetoric, however, is as much irritant as salve. 'We can't promise the earth,' murmurs the BTCV's conservation holiday brochure *The Natural Break*, 'but we'll give you a small portion to look after.' This, it makes clear, will consist of 'rolling hills, slow-flowing rivers, ancient woodlands, empty moorlands, booming bitterns ... *the sounds and sights of the real Britain*' (author's italics).

This, fundamentally, is the rationale of green belts, national parks, 'open background' farmland and much else that is integral to post-war land-use. It is the refrain of the most influential modern critiques of planning and environmental design from *Britain and the Beast* through Ian Nairn's *Outrage* to Nan Fairbrother in 1970. But its logic is that since reality is always 'elsewhere' the search for it must always be frustrating, for at the end of the day the searcher must return to his unreal half-life at home, where all that rolls and booms is the traffic and the emptiness is that of pedestrian precincts and concrete subways.

The nineteenth century had stolen beauty and morality from the cities. The twentieth now took its own peculiar notion of reality and handed this over to the countryside too, together with some of its most inspired fantasies. The effect is that of a fragmentation of human identity, the bits and pieces scattering over the landscape, but with fewer and fewer remaining in cities. The chief of these, perhaps, was a grudging and tawdry version of necessity, with little joy or optimism – the perspective of the half-awake commuter contemplating his journey to work in the grey dawn of a wet November Monday.

Hence the sense of wonder and mystification which greeted the discovery of Moseley Bog. What properly belonged elsewhere had come home, reality – 'real existence', to borrow, once again, Eliade's phrase – was a part of daily life. Martin Buber, in his *Khassidischen Bücher*, tells the story of the rabbi who dreamt of wealth in a far-off place, journeyed many miles to find it and was then told that it lay buried behind the stove in a corner of his own house.

The parable, like that of Moseley Bog, concerns riches found where they are least expected: at home. Shortly after the campaign to save the bog was launched, it was designated a site of special scientific interest (SSSI) by the Nature Conservancy Council. Ostensibly this was because of its biological diversity and rarity value. SSSIs are curious creatures, however. There are good reasons for describing them as camouflage: a way of giving scientific, and thus political, respectability to simple human delight in natural beauty. Ecology thus again reveals itself as a heterodox science, a form of behavioural adaptation by which the imaginative and the spiritual – qualities now subsumed under the general heading 'subjective' – attempt to ensure their own survival in a deeply 'objective' culture. The campaign leaflet was more explicit. It said, 'We all need to have the experience of being in a wilderness . . . .. This site is more than an SSSI. It's magic.'

The story of Moseley Bog is far from unique. As the following pages show, there are hundreds, indeed thousands, of cases from the 1970s onwards where something similar occurred. Sometimes it was, like the bog, a physical piece of forgotten country. More often, it was a purely inner landscape, a mental picture gallery where each scene embodied some remembered delight or half-forgotten need, which was suddenly capable of being projected into the real world: on to the blank spaces of wasteland opened up by the emptying cities. In this sense, the cognitive emptiness of the moors, countryside and farmland beyond the green belt was now appearing physically in the cities. Urban decay was thus the opportunity to create cities, literally, in a different image.

The picture that emerges is thus one of discovery, of an urban society beginning to look at its immediate surroundings with fresh eyes, seeing new possibilities in old things. A radical change in perception is involved, similar in many respects to the cultural watershed of the late eighteenth century when wild scenery – the Lake District in particular – was suddenly 'sublime', having only recently been frightful, barbarous and chaotic. Yet this revaluation, and the piecemeal, small-scale reconstruction of the cities which it prompted – so piecemeal at times as to be invisible to the casual observer – could not have happened before it did because many ingredients were involved. All had to come together at the right place.

How this happened will be examined in more detail in the next chapter. Before that, it is worth completing the parable of Moseley Bog to show how, from small beginnings, revolutions can be made. In July 1980, less than two months after the launching of the Moseley Bog campaign, a small group of local naturalists, planners, teachers and landscape architects joined forces to form the West Midlands Urban Wildlife Group, an avowedly activist body dedicated to a greener Birmingham. One of its founders was Chris Baines, self-confessed 'hack' landscape architect turned wild garden propagandist, whose philosophy was deceptively simple. If urban greenspace could be used to sustain meadow flowers, skylarks, dragonflies and hedgehogs – to provide a 'countryside experience' – then life could be far more pleasant for people living in towns. But the UWG also fulfilled that daydream of the Moseley Bog campaigners – bringing 'experts' and 'people' together. As a voluntary organization composed initially of professionals, and hence something of a hybrid, it sought to build bridges between the 'consumers' and the 'manufacturers' of landscape. On the one hand was that mass of ratepayers and residents known, usually inaccurately, as a 'community'. On the other were the councils, public authorities, land managers and developers. Chris Baines and his friends wanted to involve the former and influence the latter.

By 1984 the group had 350 full members, 3,000 life supporters and up to eighty staff, the latter funded largely by the Manpower Services Commission. It had staged a Wild West Midlands week, a Dawn Chorus Day and a daisy-chain competition, and was helping to manage three of Birmingham's ten nature reserves. Its basic practical task of surveying, restoring and creating greenspace in the West Midlands involved its own planning and ecology teams and a twenty-eight-strong landscaping unit. It was advising on 100 sites each year, while its campaign list read like a gazetteer of a forgotten Birmingham: Snow Hill, Plantsbrook, Woodgate Valley, Moat Farm . . .. In effect, it was a free-floating state-funded organ of aid, advice and environmental action, combining voluntary commitment with professional skills and experimenting with newer forms of democracy, notably the making of a neighbourhood by the people who lived in it. It was by this time rather grandly styling itself an 'active and effective movement for social and environmental reform'. By most measures of success, it appeared to have arrived.

So, too, had its message. Convince a gardener, said Chris Baines, and you save a quarter of an acre for Britain's wildlife: convince a

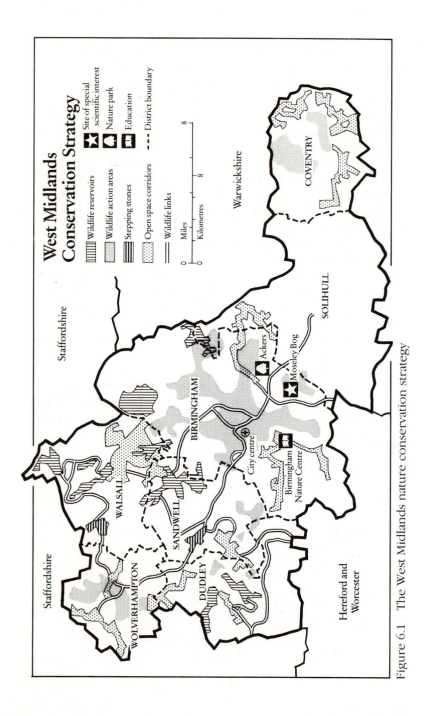

Figure 6.1   The West Midlands nature conservation strategy

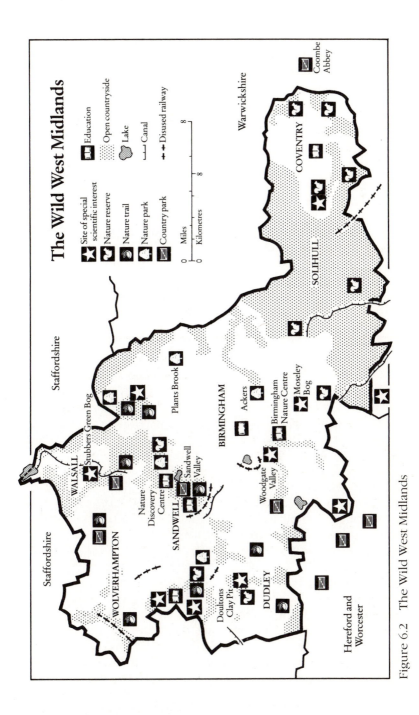

Figure 6.2 The Wild West Midlands

parks superintendent or a civil engineer and you save hundreds, thousands, more. In 1984 a landmark was reached in the greening of the cities when all the local authorities in the West Midlands – seven district councils, one county council – agreed on a comprehensive nature conservation strategy, the first of its kind in the country.

The content of this remarkable document – its designation of inner-city wildlife action areas, of green reservoirs, corridors, stepping-stones, its plans for hedges, flower meadows, 'wild areas', in parks and on road verges – is in a sense less important than the vision it expresses.

Conservation, said the strategy, was not merely about the survival of wildlife but 'the survival of the physical and mental well-being of the human species'. People needed woods, heaths, pools and meadows, but they needed them near at hand – within a thousand metres of where they lived. A greener inner city might even persuade people to resist the lure of the suburbs and the shires, might come to be regarded as part of their 'standard of living'.

Those were the words. When describing land-use and landscapes, however, people think mainly in maps and pictures, and it was in the map of the new 'Wild West Midlands' that the vision was most dramatically captured. This showed a conurbation, not walled off from the surrounding countryside, but connecting up to it through a network of winding green threads, each thread occasionally wandering into some broader expanse of urban green before finally merging with the 'open background' beyond. The city, a pattern in green, part patchwork, part tracery, was in some strange way being turned inside-out: the green belt, complementarily, was being turned outside-in. In the eighteenth century, Pope described the new landscape movement as 'calling in the country'. The West Midlands phenomenon might equally be labelled 'calling out the city'.

Not the least unusual aspect of the new, wild Birmingham was its endorsement by officialdom. For the parable of the bog, finally, takes us back to its beginnings – to Coldbath Brook, the Happy Valley, Sarehole Mill, the River Cole, the scenes, that is, of Tolkien's childhood. The corn-mill, newly restored, is now a visitor centre, a 'living museum'. The River Cole is the subject of Project Kingfisher, a collaboration between councils, water authority and volunteers to transform a bleak and lifeless valley, dominated by large tracts of mown grassland and sports fields, into a thriving wildlife corridor. Moseley Bog, meanwhile, bordered on the east by the newly designated seven-mile Cole Valley green corridor, has a wildlife action area half a mile to the west. It would perhaps be too much to say that the queer old willow-haunted Withywindle, scourge of hobbits and spur to protestors, was being recreated as a matter of public policy and edict in the heart of a modern city. But something very like that had happened.

THE COUNTRY COMES TO TOWN

Mrs Webb's service is intended to form the basis of a conservation service which anyone can adapt to suit their own particular needs. When it was held in her own church during the summer, she decided to use a recording of whale songs in place of an anthem.

(From Mrs Webb's Conservation Service, The Second Conservation Annual, *The Conservation Foundation, 1983)*

Even during Bodkin's childhood the cities had been beleaguered citadels, hemmed in by enormous dykes and disintegrated by panic and despair, reluctant Venices to their marriage with the sea. Their charm and beauty lay precisely in their emptiness, in the strange junction of two extremes of nature, like a discarded crown overgrown by wild orchids.

(From The Drowned World, *J.G. Ballard, 1978)*

If people were deserting the cities in ever greater numbers from the 1960s, wildlife appeared to be moving in the opposite direction. Built-up areas, conventionally characterized as urban deserts, were discovered to be thronging with foxes, frogs, weasels and dormice. Kestrels nested over the Strand and Fleet Street, black redstarts above Westminster tube station. Peregrine falcons hunted from the Post Office Tower in Swansea, painted lady butterflies haunted derelict Seaforth docks in Liverpool, a colony of morels – an exotic form of mushroom – turned up on the track of the disused Snowhill railway station in Birmingham city centre. In the winter of 1982 pheasants, woodcock and a short-eared owl were sighted, respectively, in the City of London, Holborn and the Strand. Foxes were variously reported as drinking from puddles in Horse Guards Parade, and fleeing a pursuing police car in Trafalgar Square. The growing areas of urban wasteland meanwhile bloomed with garden escapes like blue buddleia, goldenrod, Michaelmas daisies, or with more traditional bomb-site flora like yellow Oxford ragwort and the tall bright purple spikes of rosebay willow herb, or 'fireweed'. Such wasteland received a new designation, 'urban commons'.

The urban commons are part of the 'unofficial countryside', the striking phrase coined by Richard Mabey in 1973 to describe sewage-works, gravel pits, old quarries: places where nature, like Moseley Bog, ought not to be but keeps on turning up. The city, it was found, was surprisingly rich in such unofficial countryside: cemeteries, hospital grounds, old churchyards and railway embankments, but above all the wasteland left by the exodus of people and jobs.

These unofficial riches should perhaps have surprised nobody. Almost a fifth – 18 per cent – of towns and cities is open space. Much more, however – one study in Brussels suggested up to 50 per cent –

is actually green and busily photosynthesizing. Houses, most with gardens, make up 54 per cent of the main built-up area, which is thus far less built-up than most people imagine. And even buildings can serve as home to bats, birds, owls and a small army of lesser flora and fauna. That people continue to be surprised is evidence, yet again, that many of us may *look* but we do not necessarily *see*. Ironically, the official urban countryside of municipal park and garden was, by comparison, sterile and monotonous. 'Nature' had given way to railed-off ornamentalism, wildlife had been chopped, sprayed and mown almost out of existence. 'Green deserts' surfaced with 'soft concrete' and spotted with lollipop trees: this was how the new breed of urban green characterized such treasured London landscapes as the royal parks.

They were making a judgment that was aesthetic, spiritual and also ecological. The urban greening movement could not begin until ecology had supplied it with an intellectual framework and environmentalism with a sense of loss and outrage that groped towards a spiritual vision. Nor could it begin until a place was found and a lesson learned.

For a culture steeped in rural nostalgia, the lesson was particularly traumatic. It is summed up in the contrasting words of Nan Fairbrother in 1970 and Marion Shoard a decade later – ten years which seem to span two different worlds. 'Our new style farmland,' wrote Miss Fairbrother, 'is a living and thriving landscape efficiently adapted to the modern world and productive as it has never been before ... though we may regret the old countryside which it is ploughing away, we must still admire its vitality.' And there was always, she added, with a touch of unintended humour, the 'unspoilable sky'.

For Marion Shoard in 1980 the issue was brutally clear and the message was quite different. Paradise itself was under threat. A 'sentence of death' was being carried out on the English landscape. 'The executioner is not the industrialist or the property speculator .... Instead it is the figure traditionally viewed as the custodian of the rural scene – the farmer.'

The mournful litany of statistics on grubbed-up hedgerows, uprooted woods, ploughed meadows and drained wetlands, is by now familiar. Miss Fairbrother might enthuse about the 'unlooked-for new pleasure of the new open landscape' but the urban armies of the fourth wave found, increasingly, characteristics they thought they had briefly escaped – a precision-tooled terrain with the scale, uniformity and neatness favoured by the civil engineer and the municipal bureaucrat. The country was being remade in the image of the city: romance and refuge were eroding.

None of this is to extol sentimentality or to decry openness. Some of the most prized British landscapes – the Lake District, the cliffs of Cornwall – undoubtedly owe much of their appeal to a quality which Jay Appleton, the geographer who has elaborated one of the most compelling explanations for our feelings about landscape, calls 'prospect'. Appleton seeks to explain human behaviour in evolutionary terms. He argues that since man evolved as predator and prey on the great open savannahs of prehistoric times, he is happiest when he can see without being seen. But there remains the question: prospect of what? What *kind* of openness?

Some of the hedgerows grubbed up by the farmers – but, as Oliver Rackham has made clear, *only* some – were themselves comparatively recent in origin, a product of the eighteenth-century enclosures. The open-field system which preceded them was, as its name implies, a more open landscape – in parts. In other parts it was more closed. As the estimates of the seventeenth-century chronicler Gregory King show, there was far more forest, wild land and rough pasture, and less cropland, than today.

But it is also disingenuous to argue that since landscape tastes do change, they must be infinitely malleable – that, as Nan Fairbrother implies, we must learn to like what we have to lump. In accustoming ourselves to engineered landscapes, in both city and country, we will become different people. A price will have to be paid, just as it was when civilization learned to love wilderness by losing it. To argue otherwise is to argue that men are not changed by what they see.

Few men, for example, could exceed Thoreau in his love of peace and prospect – 'The man I meet,' he once remarked, 'is seldom so instructive as the silence which he breaks' – but even he was defeated by the bleakness of Mt Katahdin in Maine. Nature from its summit, he said, was 'vast and dreer and inhuman'. The landscapes of the new agribusiness are a step towards those of desert and prairie, in which 'things in themselves' fade and empty. The Canadian prairie has indeed been described as 'an experience, not an object – a sensation, not a view ... a way of being ... not a thing at all'. John Berger has written similarly of Miss Fairbrother's unspoilable sky. The sky has, he says, 'no surface and is intangible: the sky cannot be turned into a thing or given a quantity.'

'Such landscapes, in other words, invite reverie and resist

engagement. The spirit at home in them is one which, like Thoreau's, prefers isolation and self-absorption to the speech of others – a far cry, indeed, from the sociable bustle of the eighteenth-century pleasure garden. Outrage at the march of agribusiness thus becomes an insistence, however confused, on certain values – intimacy, smallness of scale, sociability, even perhaps 'community' – as well as the need to re–establish personal connections – touching, feeling, smelling – with that remote abstraction known as nature.

This was the vision of environmentalism, a vision of real existence coursing through the green world, energizing people and plants alike. Contact – daily, direct, renewable, not carborne on bank holidays – had to be made with it. But a means of contact – a language, a programme, above all a philosophy of things in themselves – had first to be established. As Berger comments, an oil painting defines the real as 'that which you can put your hands on'. It is difficult to connect with the sky.

Ecology could supply this but it had first to cure itself of the prevailing presbyopia, that optical condition, the opposite of myopia, in which only objects in the far distance can be seen clearly. Ecologists, like everybody else, preferred 'natural' sites in the countryside. Charles Elton, one of the founders of animal ecology, somewhat grudgingly included the 'domestic system' of buildings, parks and gardens in his habitat classification of 1954 but went on to describe it as 'partly impoverished [and] partly peculiar'. With one or two well-rehearsed exceptions – the flora of bomb-sites, the impact of air pollution – ecology paid the unofficial countryside little attention until the later 1960s.

In the 1970s urban ecology developed into a recognizably separate discipline. Yet this only mirrored a widening recognition not merely that nature had survived in cities but that it possessed its own intrinsic and distinctive value. It was not, in other words, a shoddy, second-best imitation of real countryside but *reality itself*, or at least a legitimate version of it.

The fascination with the flora of dereliction provides one example. The very process of neglect and urban decay led to rampant growth, the slow effacement of prescribed identities, the emergence of authentically original landscapes which offered their own atmos-phere, their sense of place, and hence a renewed connection with surroundings usually perceived as devoid of identity. As forgotten wastelands turned green, each city developed its own distinctive urban common. Bristol's was dominated by naturalized wild fig and escaped buddleia, Manchester's by tall clumps of late-flowering knotweed, Sheffield's by a riot of garden escapes, flowering throughout the summer, from feverfew and goat's rue in June to Michaelmas daisies in October.

Ecological succession, the principle that plant communities progress unchecked through several different stages to reach a 'climax', a stable community with a dominant species, meanwhile produced a complex range of regressing landscapes. Unused concrete filter beds built by the water company in Hackney, east London, by the River Lea have become reedswamp and marsh. The Welsh Harp reservoir, made by the Victorians to provide water for the Regent's Canal, is now largely fen – or, more accurately, a novel, man-induced habitat category of fen-with-reservoir. The most poignant instance of such natural alchemy was the transmutation of London's great Victorian cemeteries – Highgate, Tower Hamlets, Nunhead – into overgrown Gothic-Italianate fantasies where ruined chapels and ransacked mausolea were simply overwhelmed by a tidal wave of green.

These examples, and many more, taught an important lesson. As the landscape architect Jenny Cox said in her report on Highgate Cemetery in 1976, 'Nature has more imagination than man and will produce more organised and intriguing patterns of her own accord than we will ever achieve.' She was speaking of the intricate disposition of ivy on the cemetery's 51,000 tombs, but it was a lesson that was taken as applying across the range of plant, animal and bird species, even to the microfauna of the soil. Man simplified: nature made more complex. The lesson was ecological but it also expressed social and political values since in complexity lay diversity and hence, prospectively, greater freedom of choice.

The model of the urban ecosystem which evolved over the same period, and owes as much to developments in physical geography as to ecology, also envisaged the city as a highly distinctive addition to an expanding catalogue of natural habitats. A large city is a place, relatively speaking, of warmth, dryness and storms. There are fewer breezes but much local air turbulence. Less snow falls and less sunshine penetrates, but there is more fog, fifty times more dust and twenty-five times more gases. If any one feature stands out, it is that of aridity: the absence of water, interestingly, is a haunting theme of T.S. Eliot's urban threnody, *The Waste Land*. Cities have, for example, been likened to the parched and denuded 'karst' limestone topography of Yugloslavia – city sewers performing the functions of underground caves. Dew, dependent on plants, is sparse, groundwater tables are lowered and streams' base flow can be reduced by three-quarters through urbanization, irrespective of whether they are culverted or concreted over.

The urban 'heat island' effect, caused by the huddling together of artificial heating sources coupled with the albedo and conductivity properties of building surfaces, results in cities being on average 2°C warmer than surrounding countryside – in climatic terms, a substantial difference. At night the margin can increase to 9°C. Infra-red measurements have shown that asphalt streets and car parks can reach temperatures 20°C higher than that of the ambient air.

Cities' warmth and their jagged relief thus generate storms and extra rainfall – they have the same effect on the weather as a small or medium-sized depression – and the 'quenching' of the urban fabric is typically rapid and violent. Because of hard surfaces and rapid drainage, run-off and flooding is almost five times greater than in the countryside, where soil and plants act as a sponge soaking up moisture. Yet the 'canyon' effect of tall buildings means that dust and dirt, often toxic, accumulates in large quantities in city streets. Since wind speeds are up to a third less, the breezes simply cannot flush out the pollutants.

The new emphasis on the climate and geomorphology of cities is a recognition, long overdue, that 'landscape' does not end where cities begin. By some strange chemistry of perception, the city is thus transformed, grows instinct with new possibilities, often stark and violent. The colonizing ragwort, for example, is a plant of volcanic terrain. Office blocks form artificial gorges haunted by birds of cliff and sea cave – the tamed rock dove, or feral pigeon. The house sparrow is a species of the prairie. Each man-made change, from the clearance of buildings to the hoeing of a vegetable patch, creates an ecological niche, an opportunity for new plants to take root and thrive. A colony of tropical knotgrass discovered by a towpath in Hackney may have prospered because of canal pollution from nearby factories, creating a tropic in miniature. The Mediterranean bladder-snail has colonized the Spark Brook in Birmingham thanks to a similar microclimate, created by warm condenser water from a wire-rope works. The circumstances of both immigrants' arrivals are unknown.

In cases like these, symbiosis – the evolutionary association of quite different organisms – assumes a new potential. In the form of symbiosis known as mutualism, different species co-operate and yet maintain their separate identities. Where identities blur, however, domestication results, and the wild animal – the hostile, unpredictable, fantastic creature of *terra incognita*, of the elsewhere schema – is absorbed into the human home or *oecumene*, a type of settlement clearly marked off from the land beyond.

Domestication, for its part, usually involves a type of animal behaviour known as tameness yet the behaviour of supposedly wild animals in cities highlights the inadequacies of our vocabulary. Foxes which raid dustbins and squirrels which feed from children's hands are displaying behaviour patterns which fall into neither of these categories. They share men's home but remain unsubjugated. They profit from the human connection but do not rely on it. For the human beings, meanwhile, the feelings are complex and run deep. In the pleasure recorded by those who unexpectedly cross paths with a bold red fox in the heart of the city there are many strands: simple wonder at the event and at the creature itself; a sense of reconnection with a lost and mysterious world; and also a peculiar kind of relief, the relieved loneliness of one who finds company in a place where he had long ceased to expect it. In such encounters a new means of co-existence between men and animals, and by extension between men and nature, is being mapped out.

Cities, particularly as they have begun to unravel, have afforded more opportunities for such mutualism to develop than any other kind of habitat. Wild creatures – birds like the song thrush, greenfinch and robin (but not, interestingly, the tits or the chaffinch) – are actually 'tamer' in towns and this may help to provide the key to unlocking the ancient tension between home and the wild, in all its many forms.

Modern studies of aboriginal peoples indicate similar forces at work. The Mbuti, pygmy hunter-gatherers living in the dense forests of the upper Congo, view the forest as 'generous and friendly', personifying it as mother, father, friend or lover. The adjacent Bantu agriculturalists, however, consider it mean and hostile. Only for the white man, remarked Chief Luther Standing Bear of the Oglala Sioux Indians, was nature a wilderness, infested with wild animals and savage people. 'To us,' he added, 'it was tame.' But since the chief did not acknowledge the concept of wilderness, it is likely we, equally, cannot as yet begin to understand what he meant by tameness – beyond, perhaps, the fact that all nature was home to the Sioux.

This does not imply an ecological endorsement of the city but it does show that the phenomenon hazily known as urbanization need no longer destroy wildlife. Pollution of air and water usually harms, but can also enrich. As every gardener knows, the urban 'heat island' shelters more sensitive plants. Heathers and azaleas thrive on the acidity of city rainfall. Moreover, the attenuated city of the late twentieth century, emptier and hence more thinly spread, is, with its gardens, an increasingly potent generator of edge-zones, the ecologists' term for the boundary between two different habitat types which is thus exceptionally rich in wildlife.

The contrast has thus grown even more marked with a countryside farmed in obedience to standards which the city has begun to abandon. Urbanization around the Brent reservoir in Middlesex led, in the 140 years after it was built in 1833, to a decrease from seventy to forty-three in the number of bird species regularly breeding there. The county of Middlesex itself lost seventy-eight native species of plant in the century after Trimen and Dyer's *Flora of Middlesex* was published in 1869. Yet the county *gained* about 100 other plant species in the same period, eighty-five of them garden escapes, like goat's rue, slender speedwell, the striking large-flowered evening primrose and the better known buddleia, daisies and goldenrod. Is it richer or poorer in consequence?

Objectively, as the ecologist Brian Davis has pointed out, the question is unanswerable. Subjectively, a loss of connection with the past is involved. Yet the land taken for much of that Victorian and early twentieth-century urbanization was quite different from the land which would now be taken for development. Of the seventy-eight species lost, seventeen were associated with bogs, fifteen with heaths and moorland, others with woods, meadows, cornfields – the landscapes, in other words, that modern intensive farming has so devastated. And since variety of habitat produces diversity of wildlife, taking a typical slice of modern farmland and turning it into a housing estate or a gravel pit would probably produce a genuinely richer array of wildlife.

To say that the wildlife of the countryside has been driven into the cities by the farmer is a telescoped version of the truth that nevertheless helps to describe what actually happened. In some cases nature survived while the city grew around it. In other cases cleaner cities proved decisive. The return in 1983 of the salmon to the Thames – so filthy in Victorian times that bargemen would set fire to it, by lighting the methane produced by rotting sewage – is perhaps the best-known example, but there are many others. In central London the return of house martins and swifts has been prompted by an increase in insect food supplies resulting from reduced smoke pollution.

Other animals, however, have simply walked in and taken up residence. Foxes, according to the London Wildlife Trust's Foxwatch survey in 1983, probably came in from the counties along railway embankments. And just as ragwort escaped by train from the Oxford botanic gardens, many native plants have also travelled into cities beyond their customary rural habitat by courtesy of British Rail.

In 1978 the accumulating evidence about the unexpected riches of city wildlife was confirmed in striking fashion. Three years earlier the Nature Conservancy Council had conceded the growing significance of urban ecology by despatching the naturalist Bunny Teagle to the West Midlands with a mission to explore. To the NCC's evident surprise, he returned with many tales of marvels – of Muntjac deer and Cetti's warbler; of wild plants with talismanic names, like mignonette and traveller's joy, found in car parks and roadside wastes; of feral cat colonies living in hospital hot-air ducts; of as many as eighty species of moth trapped on the roof of the City of Birmingham Museum. Kestrels had taken kitchen scraps in Selly Oak and Hall Green, kite-like behaviour not previously recorded in ornithological literature, heather was growing again on raw colliery waste at Stubbers Green, kingfishers were firmly established in south Birmingham.

There were more depressing findings. The average park lake was effectively a 'sterile, vertical-edged concrete tank'. Rigid council mowing regimes cut back undergrowth so that it was unable even to provide nesting sites for robins, dunnocks or wrens. 'The average British municipal park,' Teagle concluded, 'is barren compared with many of the so-called derelict areas of the Black Country.' Teagle's study took its name from the 1843 Report of the Midland Mining Commission which remarked that the traveller to the area 'appears never to get out of an interminable village'. It was called *The Endless Village*.

The mid-1970s saw the Countryside Commission starting to nibble at cities, most notably in the urban fringes of London and Manchester. But the voluntary sector's response was the more significant. On Saturday, 3 May 1980, nearly a thousand people gathered in the Great Hall of Bristol University for the launching of the Avon Wildlife Trust. In the following six months, 1,500 joined, nearly a sixth of the membership of a long-established rural naturalists' trust like that of Essex. Bristol, reputedly home to more foxes (fifteen) per square mile than any other city, set the pattern for a network of similar groups springing up in the other major cities, most conspicuously the following year in Birmingham and London. In 1982 the British Trust for Conservation Volunteers conceded the existence of the real Britain in the cities as far as to appoint urban field officers in the West Midlands and Newcastle, and immediately reported an overwhelming response, with a fourfold increase in the amount of voluntary conservation work. In 1984 the Fairbrother Group was set up to act as an umbrella body for urban conservation groups which by 1985 were active in over thirty towns and cities.

Extending the conservation franchise was one of several distinctive notes struck by the new breed of urban wildlife organization. London's wildlife, after all, had been patiently and painstakingly scrutinized for over a century by the London Natural History Society, which published in *The London Naturalist* regular accounts of many of the finds and sightings seized upon as such novelties in the 1970s. This was a different, older tradition, however, the tradition of the disinterested and fundamentally passive amateur, treating his discoveries as happy accidents over which he has little control. The new groups were altogether more radical and activist, intent – passionately so, in many cases – on involvement and contact, impatient with delays and bureaucracy.

If such a fragmented movement could be said to have a collective manifesto, it was this: that people living in the cities had a need for, and a right to, their own local patch of countryside and that it should belong to them, not to the chairman of the council amenities committee or even to some distant organ of the nature conservation establishment. Where that patch of urban countryside was railed off it should be freed, where it was threatened it should be defended, where it did not exist it must be created. 'What the urban conservation movement really has to get across,' said Chris Rose, the LWT's first conservation officer, 'is how people can *use* the planning system, so that they can become better able to control that piece of land at the end of the road.' In a sense, it was unprecedented, the first attempt at practical popular conservation – conservation *for* the people where most of them actually lived, in cities.

It was, of course, a recipe for conflict, often with the well-marshalled ranks of the well-meaning. On 5 May 1981, the London Wildlife Trust had prepared its Primrose Hill declaration, a ringing denunciation of the imprisonment of London's natural heritage, an affirmation that nature and wildlife would 'reclaim yet the smallest open space on the street or window ledges'. To accompany the declaration, and inaugurate the trust, a primrose was to have been planted on Primrose Hill. Officials of the Department of the Environment's royal parks department refused permission.

The incident was a cameo of many larger controversies up and down the country. In Glasgow, an action group was formed in defence of seventy-six acres of Clydeside wasteland, alive with stoat, weasel, willow, bulrush, heron and barn owl, known as the Cunnigar Loop, 'the only bit of countryside we've got', according to residents, which the council wanted to turn into a public park. In Chiswick, west London, after a sustained protest campaign and a four-day planning enquiry, residents defeated a British Rail proposal to sell the Gunnersbury Triangle – six acres of once-rural hinterland trapped a century earlier between three railway lines – for industry and warehousing. 'The public,' said a bemused spokesman for Lovell Construction, the potential buyers, 'have been ill-informed. There are no specimen trees on site: it is a natural uncontrolled growth of low quality.'

The battle for the countryside, as these and many more examples indicate, was beginning to be fought and won in the cities. Urban conservationists had more weaponry than their rural counterparts: planning controls, resident professional expertise, the absence of an elected farm lobby as in the shires – above all, perhaps, the capacity for year-round vigilance. Typical was the wildlife watchdog scheme conceived in Birmingham, aimed not only at unfriendly developers but friends and neighbours – the man at number 24, as the Urban Wildlife Group put it, who 'may be thinking of chopping down that huge friendly old oak tree in order to construct a run-in for his car'.

Urban naturalists thus forsook their country weekends for exhaustive surveys of city wildlife and its habitats. In London subjects ranged from foxes, hedgehogs and owls through frogs, newts and slow-worms to acorns. The surveys provided a unique opportunity for professionals, amateurs, schoolchildren and pensioners to co-operate. In London and Birmingham detailed dossiers of metropolitan wildlife habitats were incorporated for the first time into local authority planning files. In the autumn of 1982 the first-ever register of protected or designated wildlife sites in the capital was drawn up. The list was by no means comprehensive but numbered sixty-eight in all, thirty of them sites of special scientific interest or local nature reserves. Of the national total of nine statutory local reserves in urban areas in March 1983, eight had been declared since 1974.

Plate 7.1   A trapped wilderness: Gunnersbury Triangle, west London, looking towards Hampstead Heath and the city centre

A new image of an old and hidden city began to emerge. As part of London's Waterside campaign, 290 miles of the Thames and its tributaries were measured, along with 3,000 acres of reservoir, lake and pond. Suddenly, wrote Melanie Roots, the LWT's development officer, London was 'nothing but water . . . a trellis of buried streams and abandoned canals'. Oxleas Wood in Shooter's Hill, Greenwich, home to the prized wild service tree but threatened with a four-lane extension of the M11 motorway, was discovered to be a vital link in the so-called Green Chain, a swathe of open land, woods and gardens across south-east London from Thamesmead in the east to Beckenham in the west.

As the significance of the viatical habitats and wildlife corridors became clear, they received new designations. An old railway track – and by extension a footpath, a bridleway, a cycle-track – was a greenway. A river or canal, somewhat optimistically, was a blueway. People began to see different moods and aspects of the city: not tarmac and concrete but woods and water and trees. In 1984 the Think Green campaign was launched by Dartington Institute and the Inter-Action Trust, initially in the West Midlands and Liverpool, where it was marked by a major international congress on green cities. The campaign title, proposing a new awareness of the urban green so that 'we can begin to like, even love our towns and cities again', neatly encapsulated the mental revaluation taking place. In part it was a rediscovery of the topography and landforms, the intimate physical relief, of a pre-industrial city. Many old farm and village ponds in the south London boroughs, some still providing a breeding ground for the internationally rare great crested newt despite tipping or partial drainage, were found. Ten were earmarked by the LWT for restoration. Active conservation thus became a home-based activity, as volunteers took up spade and sickle to clear rubbish, dig ponds and lay footpaths. Many trusts and other groups took on the job of managing or creating their own reserves, supplying wardens as well as workers. In 1981 Highgate Cemetery was bought by its Friends, numbering over 2,000 three years later, for £50. Plans for habitat creation were drawn up, for wildlife reserves and nature trails.

New intimate relations were being forged. Late in 1978 the Nature Conservancy Council asked adults and teenagers from four urban conservation projects – Birmingham, London, Swansea and Mexborough, near Sheffield – why they enjoyed the experience of nature in cities. The response was complex. They spoke of escape, freedom, adventure, discovery. They spoke too, unconsciously echoing Jung and Otto, of 'timelessness', or rather of that privileged, magical time of myth and legend, using terms like 'paradise' and 'oasis'. But an especially poignant note was struck by the sense of contact and identity, both with nature and with other people. 'Fun with dirt' was how one Birmingham schoolboy expressed it. Others dwelt on the rediscovered richness of ordinary sensations: fresh air, the 'feel' of flowers, the crackle of breaking ice, above all the smells, 'smells,' as one Londoner said, 'you wouldn't smell anywhere else. Your whole senses are alive.'

Strong forces, spiritual and psychological, underlay this renewal of contact. They were hinted at in the tableau of 'sacred and historical trees' staged by a theatre workshop at One Tree Hill, in Honor Oak, south London, and in the 'tree hug' by children in Oxleas Wood in late 1983, a form of ceremonial goodbye to some of London's longest-lived oaks and wild service trees. In India the Chipko movement signals its dependence on trees for sustenance in precisely the same way.

Woods like Oxleas, or Sydenham Hill, in south London, also demonstrated ecology's growing grip on minds. Wild service tree

Plate 7.2　The green desert: empty grasslands in Dulwich Park, south London

and wild cherry, both found in Oxleas, are indicator species: they grew in the ancient woodland that covered Britain in the wake of the retreating ice sheets some 11,000 years ago. Oxleas and Sydenham Hill are remnants of this. They are thus survivals from a purer, more authentic and elemental world, existing in an age before agriculture, cities and civilization.

That, at least, is a generic, 'emotional' response to old landscapes: that they are pristine, 'timeless'. Ecology's achievement has been to take this essentially private, uncommunicable experience and devise for it, in categories such as ancient woodland, a visible, living, public presence. In so doing it has provided a symbolic language, a picture of a Britain once clothed in forest, that can be shared with others, can inspire community action. In this shared landscape of the imagination, a far more accurate blueprint of the future than the merely physical forms of stone, wood and water, the ancient woodlands survive, grow and merge, permeating the cities, ulti-mately – possibly – overwhelming them.

This was the vision of Richard Jefferies and it is the vision of a body of fiction from the Flood and the Apocalypse to the genre of novel and film so popular since the later 1960s depicting humans alone in a world after holocaust. The common theme is that of a civilization fallen prey to pride and arrogance, but the human response may vary from the experience of melancholy at a moss-encrusted ruin – Shelley's 'Ozymandias', for example – to a thrilled and dreadful fascination culminating, as with Jefferies's sinking of London beneath the swamp, in a disguised willing of the holocaust itself. Martha Wolfenstein, as we have seen, called the phenomenon 'post-disaster utopia'. Nuclear war or eco-catastrophe, the modern mind appears to be saying, may be the quickest way of having the world to yourself again.

There are slower ways, however. The reinvention of the ancient wood in the city, the rediscovery there of swamp, marsh and fen, the elaboration of green chains and blue ways, all suggest that post-disaster utopia is for the first time forming the basis of a popular programme of urban reconstruction. They also prove that the mood of black depression which fell on many Victorians when they considered the cities they built and loathed was quite misplaced. In 1898 the great naturalist W.H. Hudson declared confidently, 'It is exceedingly improbable that any of the raptorial species which formerly inhabited London – peregrine falcon, kestrel and kite – will ever return . . . .' The prediction grows more inaccurate daily.

8

A PEOPLE'S LANDSCAPE

It became green everywhere in the first spring, after London ended, so that all the country looked alike . . .. Footpaths were concealed by the second year, but roads could be traced, though as green as the sward, and were still the best for walking, because the tangled wheat and weeds, and, in the meadows, the long grass, caught the feet of those who tried to pass through . . .. By the thirtieth year there was not one single open place, the hills only excepted, where a man could walk, unless he followed the tracks of wild creatures or cut himself a path.

(Richard Jefferies, After London, *1885)*

If the country would not come to town of its own accord, it must be brought there. In the late summer of 1979, two years before the riots, a dozen small patches of wasteland in the centre of Liverpool began to assume an unfamiliar look. Banks of yellow gorse and purple heather arose. Densely planted stands of young oak, birch and willow appeared, raw and straggling, clearly experimental. Areas of suspiciously long grass took shape.

In the spring of 1980 plantings began to mature and cohere into semblances of meadow, copse, heath and hedgerow. Flowers bloomed: not bright battalions of begonias, but spindlier, less precocious specimens once known as weeds, or plants of waste and wayside. First would come black medick and ox-eye daisy, then agrimony, hawkbit and self-heal, many with the Latin prefix *vulgaris*, or common, in the botanical textbooks but clearly nothing of the kind in Merseyside. Dozens more gap-sites followed, many maturing into flowery meadows of fescue, mown only twice a year, attracting birds, insects and butterflies. In run-down inner Liverpool someone's dream of the countryside was becoming reality.

If the greening of the cities has a spiritual home, that home is a small slum clearance area of Liverpool 8, better known as Toxteth, in May 1975, when three final-year geography and geology students at Liverpool University joined forces to form the quaintly titled Rural Preservation Association. Shortly before they were due to get their degrees, the trio had found themselves sitting in a pub discussing their futures. Said Grant Luscombe, one of the three, 'We looked at the beautiful things in the country and wondered why they could not be recreated in the city.'

Luscombe and his colleagues gave their new enterprise a curiously old-fashioned name. Their aims start from the premise that many

Plate 8.1   Stages in the greening of Toxteth: a gap site left by a demolished church, later (Plate 8.2) reclaimed and planted with wild flowers and native shrubs

Plate 8.2   Stages in the greening of Toxteth: an anchor becomes
'appropriate' art – outdoor sculpture, play structure and a
reconnection with Liverpool's maritime past

inner-city problems 'stem from a kind of spiritual malnutrition caused by a distancing of people from nature'. It is a sentiment that might have been expressed by Thoreau or Olmsted, but the RPA coupled it with an ecological perspective in which people became creatures of edge-zones, congregating naturally, for example, where sea meets sand or on the fringes of a wood. Most crucially, perhaps, they came to view the emptying of the cities, which bit into Liverpool particularly deeply, as a cause for rejoicing. Urban space was an asset. It was a chance to rebuild, a golden opportunity offered to planners.

Among the first experiments was a 'community creeper' scheme. Residents of inner areas like St Michael's and Granby, many living in cramped terraced houses without a patch of garden, were offered tubs ready-planted with clematis, ivy and Russian vine simply to get them used to greenery again. Then in September 1979 the RPA launched its Greensight project: a unique example of what professionals now call the naturalistic approach to open-space landscaping.

Greensight, however, was not a professional project in the accepted sense of the word. It sought the active involvement of local people in local landscapes – and as a voluntary organization, the RPA was open to all to join and influence. Its naturalistic method was cheap and low-maintenance. It confronted urban dereliction in one of its heartlands. Perhaps most significant, it was a triumph for sheer individual initiative, drawing along an often resistant officialdom in its wake. In the oldest and most exact sense of a much-abused word, the founders, helpers and workers of the RPA were pioneers, peasant footsoldiers who clear a way in hostile terrain with spade and pick axe.

Pioneers, however, need inspiration and it is doubtful whether Greensight would have started when and as it did without the existence of a movement fusing evolution, ecology and environmentalism into a new philosophy of landscaping. This developed, in turn, and partly through sheer historical coincidence, into a practical programme of landscape creation. ·

Within landscape architecture, that body of professionals whose job it is to design 'whole' landscapes, it took the form of the so-called ecological school of design. This identified itself to a larger public in 1973 when the Landscape Research Group and the Landscape Institute held a remarkably influential symposium on nature in cities. But the greatest impact was probably that of Ian McHarg's *Design with Nature*, published in 1969, and providing a lucid and highly original formulation of the link between human and natural values. In one sense, McHarg's achievement was to capture a mood. For practical examples, the ecological movement in Britain turned increasingly to the Netherlands, where that mood was already producing, by the early 1970s, a stream of experimental landscapes designed to be played in, walked through, touched and smelled, to provide freedom and excitement. At Utrecht a reed-fringed wetland was laid out round an office block. At Delft the courtyards of high-rise flats became woodland glades with hiding places for old and young. Again at Utrecht, around new housing, logs, stones and railway sleepers were planted idiosyncratically with wild rose, willow, locust and firethorn, to create a rampant fantasy landscape for children, dense with bramble and studded with small hidden camp-sites. Living fences were planted, of nettle and thorn. Paths in new parks were not laid until people had revealed their desire-lines – shown, that is, where *they* wanted to walk. Paving stones were torn up to make way for plants in the newly pedestrianized residential streets known as *woonerven*. In the solidly built-up inner areas of The Hague old buildings were knocked down to make way for parks. Fresh types of green space were invented: cuddle gardens for children, for example, or *heemparks*, for all age groups, where native

Plate 8.3   The Dutch example: Wetlands and woodlands next to high-rise housing at Haarlem

flowers – the plants of *heem* or home – grew in their natural Dutch habitat of dyke, dune and polder.

The Dutch experience is an emphatic reminder of how a revolution in landscape values is indissolubly linked to new political, social and human goals. No other national government, remarked Brian Berry, 'demonstrates quite the same concern for the social consequences of urbanization.' Reconciliation of the 'two vast monocultures' of town and country – the need proclaimed by the Dutch writer Louis Le Roi in 1960 – provided not merely a fresh and fascinating environment for the office-workers and flat-dwellers of Delft and Utrecht but a national land-use policy. Berry calls it polycentric 'concentrated deconcentration', based on environmental goals and with the aim of preserving the open green heart of the Randstad, that great horseshoe-shaped sweep in which the Netherlands' biggest cities are arranged.

A single simple technique like desire-lines for paths in parks is similarly a product of that rejection of machine-age democracy associated with the Provos, those fantastical white-coated figures of the later 1960s. The young, said H.J. Bos, one of the pioneers of the new Dutch landscapes, wanted an alternative. 'They want to see and use green space ... they want to walk where they like and not only on paths that someone else has laid out for them.'

In some respects, it was social engineering: an attempt, late in the day, to mitigate the tower-block excesses of the early 1960s, most notably the gaunt 260-feet high Bijlmermeer estate in Amsterdam, intended for 100,000 and close in design to Le Corbusier's Ville Radieuse. The estate, like most of its contemporaries in Britain, was plagued with appalling social problems, and there is something extraordinarily touching about the solutions ultimately adopted. Set against the great straight towers of monumental concrete, against the more or less constant howling of the cold North Sea wind, were trees and flowers – oak, ash and beech, cornflowers and poppies. The experiment was studied, and a highly positive verdict delivered, by the Netherlands Institute of Preventive Medicine. Environmental determinism was exploring new frontiers.

There are other reasons which may explain why these things happened in the Netherlands. As one of the most densely populated and urbanized countries in the world, it has been directly confronted with some of the acutest problems of city living. It has also probably the longest and most continuous history of land reclamation in the world, stretching back beyond Vermuyden in the seventeenth century to the Middle Ages. Nearly half the country is man-made. Reclamation is living proof that land is recyclable, that it need not be lost forever because it has 'disappeared under concrete'. In other words, it helps people to think *ecologically*, in cycles rather than straight lines. But as an emblem of large-scale interference with nature it has always aroused deep misgivings. Descartes witnessed reclamation in the Netherlands and went on to propound his theories of man's mastery over Nature. The sixteenth-century Dutch engineer Andries Vierlingh justified it, disingenuously, it now appears to us, on the grounds that men could only build dams and reclaim land by virtue of the skills God had given them. 'Making new land,' he thus concluded safely, 'belongs to God alone.' Even deeper feelings, however, are involved when such interference takes the form of the wholesale removal of a familiar landscape, one perhaps remembered from childhood. The bitter sense of loss felt in such circumstances stems from an instinctive feeling that place – usually a *particular* place – forms part of people and people form part of a place. Anthropologists, as we have seen, call the phenomenon autochthony. Laypeople tend to call it 'roots'. The theories of Jean Piaget, the leading figure in modern developmental psychology, probably offer the best clue. The child, says Piaget, does not learn by passively receiving information from a 'real' environment, but by exploring, initiating contacts, structuring his experience – by 'acting-in-space'. Our own comings and goings 'provide the framework for our memory images of districts and landscapes.' Place and person are thus bonded together by an altogether more intimate and creative process than that implied by the western scientific tradition of observer and observed.

Piaget's experiments support the Kantian notion that the known cannot be separated from the act of knowing. Given that landscape is also a projection of mind, chiefly of memory and association, the obliteration of a familiar place strands us in the present, disorients us by refusing to yield up those old friends and moments with which the mind persists in peopling it. The connection has gone but the image remains. Where it persists we call it delusion and consult a doctor.

Familiar places disappeared by the skipful in the slum clearance and urban reconstruction of the 1950s and 1960s. By 1969, as John Barr's book *Derelict Britain* pointed out, the decline of many of the earliest smokestack industries was already generating large-scale blight, represented perhaps most dramatically by the lower Swansea valley, an urban landscape poisoned by two centuries of metal-smelting and covered in seven million tons of slag. The logic of planning controls, population and housing pressures, and, as the 1971 census results disclosed, accelerating decentralization, all pointed to an unprecedented need to reclaim land, much of it urban. But land could not be reclaimed without new landscapes being created. Nature would have to be redesigned, and on a massive scale. Whose pattern-book would be used?

The post-war years found landscape architects in a philosophical cul-de-sac. With the – occasionally outstanding – exception of the New Towns, their work was an afterthought to public design, a form of statutory prettification of some bleak road scheme or power station. The relationship between the designer and his audience had, in Britain as in the Netherlands, meanwhile assumed a new importance, particularly in the 1960s, a theme explored in the 1969 Skeffington Report on public participation in planning. For landscape architecture it was a particularly relevant theme since the two most memorable achievements of land design, the landscaped park of the eighteenth century and the city park of the nineteenth, had both been, in an important sense, experiments in liberty which ended in the denial of liberty.

Ecology offered a way out of man-made aesthetics and proprietorial landscapes. It proposed filling empty urban wastes with the 'real' landscapes of the countryside. It provided a palette of designs, or habitats, based on neutral, scientific observation: acid heath, meadow, wetland, chalk grassland, woodland edge. It also offered judgments based not on whim but on painstaking analysis which gave fresh meaning to concepts like richness and complexity and new dignity to the past.

A native tree like the oak, for example, supports 284 species of insect. Close behind comes the willow which supports 266, the birch (229), the hawthorn (149) and the blackthorn (109). By contrast sweet chestnut, introduced by the Romans, and walnut, a fifteenth-century arrival, between them provide home to only eight species. A similar story can be told of many of the exotica imported since large-scale collecting began in the eighteenth century, including favourites like rhododendrons and azaleas. The main reason is that the oak and its fellows were established before the post-glacial flooding of the English Channel cut Britain off from the Continent 8,000 years ago. Associations have had time to build up. The web of symbiotic and predator-prey relationships centred on the native oak and its companion insects is correspondingly far more comprehensive and intricate.

Planting an oak thus becomes a peculiar and noteworthy act of human affirmation, the insistence on an identity, a sense of self, far richer than our surroundings allow us. It is a way of connecting, both with the lived past and with a living present. As such, it is a symbol, but it is also, as the buzz of insects on a summer afternoon reminds us, a thing-in-itself, an embodiment of some ambivalent but vital principle of existence.

Ecology, finally, offered a clarification of that curiously opaque principle of eighteenth-century landscape design known as *genius loci*, spirit of place. A particular place, ecology says, has its own geology, climate, topography, flora and fauna. This forms, in Ian McHarg's phrase, its 'landscape identity'. But it also has a human history: man has used it as his own habitat. Its *genius loci* is the sum total of these associations. The first requirement of design is to discover them. The designer must thus cease to order, posture and theorize. Instead, he must watch and wait and understand. He must become, in effect, an applied ecologist.

The idea of a design, and thus a purpose, in nature, as Clarence Glacken has pointed out, is probably the oldest of man's concepts of the world about him. The thrust of Judaeo-Christianity was to associate it with an extreme version of transcendence. In McHarg's philosophy of design with nature, however, the concept of natural order and purpose was rescued from the wreckage of Christianity. Ecology, as McHarg expressed it, perceived the world and evolution as a creative process and hence arose 'the vast relief that we are not required to be the architects of God, and assemble this vast and complex creation'. Through ecology, men could thus escape from themselves. In formulating his theory of landscape design McHarg not only assimilated a science and articulated a mood but restated, for the twentieth century, the neglected Christian idea of the good steward and husbandman.

The ecological method of landscape design has been stigmatized as merely another version of imitative naturalism, another technique at the disposal of the designer. Such a judgment ignores its deep roots in social and cultural change. In the 1970s these began to produce a series of new landscapes described, and viewed by the public, as broadly 'natural'.

The reclamation of the Lower Swansea Valley was one example. Another was the 1,500 acres of marlhole, colliery tip, disused railway and canal at Stoke-on-Trent in the Black Country. In 1972 the landscaping firm Land Use Consultants published their 'revised planting concept' for Stoke. Old rail tracks became paths and cycleways, fringed with native trees and wild flowers. On 130 acres soured by mining waste Central Forest Park arose, with its lake and long grass. Subsided chemical dumps were transformed into the Westport Park nature reserve, a string of pools bordered by reeds and alder carr woodland.

Stoke, however, was large-scale, professional reclamation. The new landscapes that began to spring up in their hundreds throughout the cities towards the end of the decade were small-scale, made by volunteers or 'amateurs', often expressing some highly idiosyncratic *genius loci*. They were called by different names: pocket parks, nature parks,' community or neighbourhood gardens. Many had wildlife or wilderness areas, some were fully-fledged wildlife reserves. The wildlife was often miniaturized: insects, spiders and butterflies, the latter especially popular.

Most were in unlikely situations. In Reading a residents' association developed Kennetside Park out of a former rubbish tip on the

Plate 8.4   Sunnyside Gardens, a home-made pocket park in Islington, north London

riverbank. Hindus in Bolton made a roof garden. Next to Stratford station, in London's East End, a butterfly garden arose, complete with pond.

There were also more fashionable locations. On a corner of Cale Street off the King's Road, in Chelsea, tarmac and concrete were broken up to make way for a nature garden and pond, with a wilderness area of buddleia and bramble. In Edinburgh a youth group created the Abbey community garden on a derelict site next to Holyrood Palace. Many of the new green spaces also arose in areas conventionally associated with the worst urban decay: Brixton, the East End of Glasgow, the Ardoyne in Belfast.

The people responsible fell into no recognizable category beyond that of ecological opportunist. They saw ground lying ugly or idle and wanted to change it. In effect they were human colonists, rooting in the cracks and crevices of the disintegrating city like the butterfly bush and the rose-bay willow herb. Insofar as they had any central organization, this was expressed in the Wasteland Forum, set up under the auspices of the National Council for Voluntary Organizations in 1978 with twelve members. By 1982 it had grown to more than 160, a clear indication of the social response prompted by urban wasteland.

But many more people simply banded together informally, without titles, to improve their own neighbourhood. Often it was one individual who saw the opportunity and supplied most of the work and inspiration. Probably the most fascinating account of such a transformation is the story of Meanwhile Gardens in London. Meanwhile Gardens was once a four-acre strip of wreckage and concrete rubble bordering the Grand Union Canal in Paddington. There a young sculptor named Jamie McCullough had a dream in the spring of 1976. His dream was that the fly-tipped wasteland reflected the social state of the area with horrific precision but that 'it didn't need to be like that . . . shaping a piece of ground could change it'.

The message of Meanwhile Gardens was that with luck, skill and commitment, enormous local enthusiasm could be galvanized and enormous problems surmounted. But there were many other lessons – and probably the chief one was that 'shaping the ground around you is the most powerful symbol . . . for taking control of your own world . . . once you've had that experience, you'll never feel quite the same again'. But what shape to give it? First, says McCullough, ask people what their favourite childhood place was. Everyone has one – somewhere secret and magical, usually small, protected, winding or

Plate 8.5   The neighbourhood landscape: Meanwhile Gardens, Paddington, west London

cranny-like – and everyone responds to the idea of recreating it. Second, because cities are usually flat, make valleys – larger dips and smaller dents within them, where people can form groups, feel at home yet remain 'in touch' with others.

If Meanwhile Gardens was a version of countryside with a social emphasis – within it were set, at various times, skateboard and bicycle tracks, a rock-climbing wall and a small natural amphitheatre for concerts – the new style of ecological park was 'pure' countryside: an attempt, after the fashion of the Dutch *heempark*, to recreate authentic natural habitats in the heart of the city.

The first ecological park was named after William Curtis, author in 1777 of the ambitious *Flora Londinensis*. It was set up opposite the Tower of London amidst the old Bermondsey docks by a volunteer *blitzkrieg* in January 1977. The operation took five back-breaking weekends, cost £2,000 and on 17 May the William Curtis Ecological Park was officially inaugurated, cramming into its two acres no fewer than twenty management compartments, or tiny ecosystems, from sedge, sand dune, copse and cornfield to the delightfully titled 'perennial wayside weeds'.

The Ecological Parks Trust, funded by the Queen's Silver Jubilee Trust and chaired by Max Nicholson, a former director-general of Nature Conservancy, was probably closest of all the new city greening organizations to the older conservation establishment. Its platform was nevertheless self-consciously radical. William Curtis, it said, was a 'new type of open space ... designed to echo within the built-up environment pleasant features of natural landscape and wildlife of the region or country, in forms suited to reasonable use and enjoyment by people of all ages'.

The park was an exercise in environmental education and also an experiment in creative urban ecology. The city is a form of habitat man has made peculiarly his own. Could man mimic nature so cleverly that other creatures would take up residence, unaware of the illusion being practised on them? And could he adjust his own behaviour so that other forms of life entered the cities, not as inmates to be trapped or tamed, but as guests – fellow creatures sharing the same habitat? At William Curtis a tiny derelict warehousing site served as a testing ground for a new philosophy for the design of cities. A profound revaluation of man's role in nature is involved. The ecological park, for example, is part farm, part zoo, part landscaped garden. From agriculture it draws the ideas of husbandry and culture, but without exploitation. From the zoo it

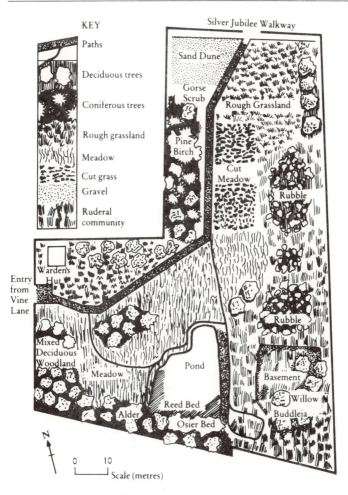

KEY

Paths

Deciduous trees

Coniferous trees

Rough grassland

Meadow

Cut grass

Gravel

Ruderal community

Silver Jubilee Walkway

Sand Dune

Gorse Scrub

Rough Grassland

Pine Birch

Cut Meadow

Rubble

Warden's Hut

Entry from Vine Lane

Rubble

Mixed Deciduous Woodland

Meadow

Pond

Basement

Willow

Alder

Reed Bed

Osier Bed

Buddleja

N

0   10

Scale (metres)

Figure 8.1   The William Curtis Ecological Park

takes men's pleasure or wonder at other creatures, but without imprisonment. In this sense it represents a further decisive step along the road from zoo to safari park, an unmistakable process of liberation. And set against the largely pictorial quality of the designed landscape, whether it is the curvilinear greenery of Capability Brown or the massed colours of Gertrude Jekyll's wild garden, is the designed ecosystem. William Curtis was a man-made landscape peopled with living things. More precisely it was a small, whole world.

The task involved was immense. At William Curtis the central lesson was that when 15,000 people, 234 species of plant, 200 of insect, twenty-eight of bird and seventeen of butterfly assemble annually with frog, toad, newt and slow-worm on two acres of ground divided into twenty mini-ecosystems, something has to give. The landscape was not as rich and complex as was first intended. Much wear and tear occurred. A full-time warden was required.

But valuable experience in ecological management was gained. Which species of *Buddleia davidii* attract most butterflies? Can water lilies, shading the surface, prevent the sunlight-fed growth of algal blooms produced by nutrient-rich tapwater? Ecological management, on which the planet's survival may depend, relies on studious and painstaking recording. At William Curtis even hedgehog droppings were noted and celebrated.

On 26 July 1985, after eight years and almost 100,000 visitors, including 25,000 Inner London schoolchildren, William Curtis ceremonially made way for offices. By then, however, it had set a precedent which many others followed. In the process the original ideas underwent a series of changes. The trust itself, in 1981 alone, was asked for its advice on twenty-three potential sites in twelve London boroughs, ranging in size from one acre in Hammersmith to sixty-three acres at a power station in Barking, Essex. It created new parks at Lavender Pond and Rotherhithe in the reclaimed Surrey Docks, began the management of a neglected wood in Dulwich, south London, and provided wardens at several other London nature parks. In Manchester an eleven-acre ecological park in the heart of the Trafford Park trading estate was planned in 1986. At Bagshot in Surrey, local groups set about establishing an ecological park because of sheer lack of public access to rural landscapes of any interest or variety. By way of the city, countryside was being reintroduced into the countryside. The test-tube landscape was nearing birth.

In Liverpool, meanwhile, the Rural Preservation Association celebrated its tenth anniversary on 21 June 1985, by changing its name to Landlife. By then it had spread to half a dozen other towns and cities, including Manchester, Leeds and Southampton, opened its own six-acre wildflower nursery, renewed sixty small pockets of dereliction, created twenty school nature reserves, and performed extensive survey work for public bodies, reporting on 113 sites in Liverpool alone. It had founded an eighteen-strong urban wildlife unit, employed 300 people on eighteen major conservation projects and seen its annual turnover grow from £900 to £250,000. Perhaps most significant, the Greensight landscapes suffered negligible vandalism, even surviving the Toxteth riots intact. The new landscapes, in Liverpool and elsewhere, spoke relatively little of human agency – that is, after all, the chief meaning of naturalistic – and, where they did, they spoke of an agency which was essentially home-grown, visible and voluntary: the neighbourhood. They were amateur in the best sense of the word: created out of affection rather than career advancement or contractual benefit. They were a sign that somebody cared, not because they were paid to care but because they wanted to.

It was from these sources that the people's landscape drew its strength and meaning. Many·of the nature parks and community gardens were conceived as temporary, 'meanwhile' – a form of provisional occupation until that particular bit of the old city represented by some heap of rubbish and builder's rubble was plugged back into place. But it is clear that on most sites the city will not return in anything like its former shape, that people do not want it to return and that the landscapes thus represent its permanent retreat across a broad front. Of the sixty-eight places listed by the London Wildlife Trust in 1982 in the first attempt at a comprehensive register of the capital's important wildlife sites, a sixth – some dozen sites – belonged to the newly created countryside of ecological park and nature garden. These were made, in the main, out of wasteland.

For these dozen sites, the listing was a sign that creative urban ecology had proved itself both ecologically and socially. These new landscapes would clearly come to be valued and defended as at present, for example, is the officially designated site of special scientific interest or the landscaped park of a stately home. In 1986 Camley Street natural park near King's Cross station became the first 'created' site to achieve the status of a local nature reserve. Yet these are democratic landscapes as the landscaped park was not and the SSSI often fails to be.

On hundreds of acres throughout Britain's cities that most powerful symbol of taking control of one's world, as Jamie McCollough called the shaping of ground, has thus become the public prerogative, not of the eighteenth-century aristocrat, the nineteenth-century philanthropist or the twentieth-century pro-fessional, but of thousands of ordinary people. Land reclamation has become a mass movement, landscape design a genuine pop art. For the nearest precedent one would have to go back to the American frontier and the colonization of wilderness by entire communities. There is thus great significance in the fact that so many people chose to make a garden. There are many possible explanations, not least the unconscious identification of gardens with ideas of rebirth and renewal, ideas rooted in the ancient paradisal symbolism of *illud tempus*, that legendary golden time of innocence and ease. Garden-making on such a scale was thus as much atavism as pragmatism, a recourse to the irrational as well as a gesture of despair with a tired city and a tired culture. But it was also, more practically, a form of creativity in which many people could join together. A modern story perhaps makes the point best of all.

In 1978 the community arts company Free Form, founded in 1969 and specializing in mural paintings, was asked by planners in Hackney for their help in the further decoration of a piece of council-landscaped open land. A mural duly appeared on the flank wall of a new estate but Free Form members were unhappy, feeling that it did not fit the site. The planners accordingly offered them a list of over 200 vacant sites and Free Form found a blank gable wall on a bare street corner at Daubeney Road, Stoke Newington, an area of new estates and old terraced housing. Discussions followed with the Clapton Park tenants' association and other nearby householders. Here were blank walls and blank land. What images should they bear? What did residents want to say about themselves and where they lived?

The answer was a garden, but a garden with a difference. It began as a dream-garden, an oddly medieval arbour-like affair with trelliswork, an aviary and conservatory, but then, where the mural touched the ground, it continued as a 'real' garden, with shrubs, rockery and a border. Artists and residents built both together, the real and the imaginary garden, with mosaic work on the mural split into jigsaw sections so that individuals could create their own element within an overall design. Free Form came to know their co-workers well: the Fords next door; Christine, who would have given anything for a full-sized tree in her garden; Wayne, aged 11, who enforced discipline among the other children and 'perhaps more than any other individual . . . had a strong sense of ownership of the site'.

In the Daubeney Road 'community arts garden', dreams became reality, art became life, design became an experiment in democracy. Other, simpler elements were also involved: fun, hard work, satisfaction, that indefinable sense of getting along nicely with others which comes from a shared cup of tea on a cold evening. These were the things the people's landscape gave back to the city in return for its unwanted wasteland. The 1984 Liverpool Garden Festival was a confused and belated recognition of this bargain, an attempt to reconstruct some of its elements on a grand and public scale.·

For the most faithful image of the people's landscape, in fact, it is necessary to return to Liverpool, not to the garden festival but to the site of the Racquets Club, that haunt of the well-to-do destroyed with such single-minded ferocity in the Toxteth riots. The most delicate wild flowers, ecologists have found, prefer a 'nutrient-poor' soil. In a rich growing medium, expensive topsoil, for example, they are simply overwhelmed by ranker competitors. To the ecologists of Greensight, therefore, it was hardly surprising, in the years that followed the 1981 riots, to find cowslips thriving on the crushed brick rubble where once the Racquets Club had stood.

9

SMALL WORLDS

Pike, three inches long, perfect
Pike in all parts, green tigering the gold.
Killers from the egg: the malevolent aged grin.
They dance on the surface among the flies.

Or move, stunned by their own grandeur,
Over a bed of emerald, silhouette
Of submarine delicacy and horror.
A hundred feet long in their world.

('Pike', Ted Hughes, Lupercal, *1960)*

I think it would be interesting to know that on my garage roof I put out some sultanas and some bread for the birds. Five sparrows came up and so did two wagtails. The wagtails ate all the sultanas but wouldn't eat the bread, and the sparrows ate all the bread but wouldn't eat the sultanas. Is this because the wagtails are berry eaters?

(Tracey Miller, aged 10, of Furzton, to Milton Keynes Junior Conservation Corps Newsletter)

The holly leaf-miner is a tiny fly of the *Phytomiza* genus with an oddly ungregarious lifestyle. The female lays her eggs in the centre of the leaf in June and shortly afterwards the tiny larva hatches out. For the next nine months the junior *Phytomiza*, burrowed mole-like into the soft green tissue below the surface, chews its way stolidly out towards the peripheries of the leafblade until it reaches its full size, about three millimetres long, in the following March. It then prepares its escape hatch, a thin triangle of tissue cut into the skin of the leaf. Here it pupates, emerging as an adult fly in late May and leaving behind it a small flap on the leaf surface.

The most unexpected feature of *Phytomiza*'s tunnelling, however, is that it appears to benefit the holly tree. Most holly leaves fall in June and July, fewer in the later summer and autumn. By placing white plastic trays beneath the trees and counting the number of mined and unmined leaves falling throughout the year, it is possible to demonstrate that almost all of the leaves falling between January and May have been mined by *Phytomiza*. The fly may thus play a vital role in ensuring that nutrients from the decaying leaves are recycled back into the tree, through its root system, at an even rate throughout the year, rather than for a limited season.

Phytomiza's molehill is a puffy, irregular,, yellowish-brown blotch or blister marring the smooth shiny green of the holly leaf. The mines are common but extremely small. This means that one has to look carefully to find them, more carefully to describe them and more carefully still to understand them. Indeed, a complete adjustment of focus is required: from the far distance to the near at hand, from the casual and coincidental to the intimately connected.

The story of the holly leaf-miner is told in Jenny Owen's *Garden Life*, published in 1983 and a small classic in the tradition of Gilbert

White's *The Natural History of Selborne*, Thoreau's *Walden* and Aldo Leopold's *A Sand County Almanac*. It differs in one important respect from its predecessors, however. In *Garden Life* fascination and marvel are discovered, not in open country, woods or wilderness, but in the landscape that lies on most people's doorstep. Jenny Owen is an ecologist who since 1971 has found in her 784 square yards of suburban garden in Leicester, two miles from the city centre, a diversity of plant life rivalling, she believes, that of the tropical rainforest. Here are moths whose names spell a threat, chart a legend, tell a joke: the snout, the shark, the ghost, the coxcomb prominent and the common footman, the chimney-sweeper, the seraphim and the pale brindled beauty. Here are toadstools and milkcaps, pink, orange, dazzling white and yellow, their hyphae intertwining and ramifying endlessly and invisibly in the soil to burgeon suddenly on the small lawn like a revelation. Here, in a miniaturized *danse macabre*, a *vespula* worker wasp and a hoverfly lock in combat, the wasp slowly eating the drone alive until, sated, it flies off, an encounter 'as violent and chilling as a lion's bloody onslaught on a zebra'.

In her Leicester garden Jenny Owen has recorded more ichneumon flies than in the much-studied Wytham Wood, near Oxford. Thirteen species of fly and wasp were new to Britain and two new to science. About a third of Britain's insect fauna can be expected to visit her garden, particularly butterflies, attracted to the flowers like motorists to a petrol station.

Jenny Owen's garden is densely planted, 'cottagey', productive of food as well as flowers. She eschews pesticides, avoids pure stands or monocultures of vegetables or grass, regards the 'clean' garden as an exercise in domination rather than management. The rich buzz of insects on a summer afternoon is living proof that food chains thrive, that energy flows and nutrient transfers prosper. But her garden is not, she insists, a wilderness. Indeed its richness would diminish if there were no gardener. It is man – or rather, woman – who creates the edges, ecotones and discontinuities of habitat. In ecological terms it is kept in a permanent state of succession, never allowed to progress towards the natural climax of woodland.

The resulting riches, Jenny Owen believes, fully entitle Britain's fifteen million gardens, covering an estimated one million acres in England and Wales alone – an area rather larger than the county of Hampshire – to be called the country's largest nature reserve. Indeed, is not the whole country, managed and manipulated, a gigantic garden? How we behave in our own garden thus reflects how we behave in the world at large. Several fascinating issues are raised by Jenny Owen's garden. The most important concerns ways of seeing. In the 1970s the wild garden was discovered, a discovery ranking with that of the flora of dereliction. In other words, eyes were opened: a fresh vision was apparent. Moreover the appeal and potential of the wild garden lay in the cities and suburbs, a further indication of the blurring of the two ancient opposites of settlement and wilderness.

The wild or natural garden has, of course, been 'discovered' before, by Pope in his riverside villa at Twickenham, by William Robinson, that powerful Victorian propagandist against bedding-out and 'pastry-work gardening', and, of course, by Gertrude Jekyll, at her Lutyens-designed home at Munstead Wood in Surrey. Set against the contemporary wild garden, however, these would now seem exercises in formalism. Pope's garden, in particular, featured a grotto inset with mirrors and a rotunda of shells, while the rhododendrons planted a century ago under the influence of Robinson and Miss Jekyll are now being strenuously uprooted where they have escaped into ancient woodlands. 'Nature', like 'city', is a work not to be taken at face value. It is the reality the word describes, whether a mirrored grotto, a rhododendron bush or a drone of insects signifying healthy food chains, that provides the chief clue to its shifting meaning.

Unlike the natural gardens of previous centuries, that urged by the new urban wildlife groups, by the chief conservation organizations and by authors like Gordon Beningfield, Ron Wilson, Michael Chinery and Chris Baines, is concerned with ecology as much as aesthetics, and with birds and animals as much as plants. In January 1984 Gardening for Wildlife Year, declared by leading conservation bodies including the Royal Society for the Protection of Birds and the Royal Society for Nature Conservation, together representing 650,000 members, was launched with Chris Baines's BBC2 television programme 'Blue-tits and Bumblebees'. The response was impressive: 40,000 letters from viewers. Sixteen months later a further defeat for horticultural convention was registered when Baines, with the help of the British Trust for Conservation Volunteers, built a wildlife garden, measuring only ten yards by fifteen yards, for the 1985 Chelsea Flower Show.

The contemporary wild gardener is the warden of his own wildlife reserve, the guardian of a secular sanctuary. He provides a home for his fellow creatures – boxes for birds or bats, ponds for frogs and newts, refuges of stone or wood for hedgehogs and toads. He leaves logs and tree stumps to rot, to encourage spiders, mites and fungi. He sets aside some odd corner for benign neglect, in the hope of some small miracle like the burgeoning of bright toadstools one unexpected autumn morning. And if the filamentous hyphae of the toadstools, endlessly feeding and growing out of sight underground, are richly suggestive of immanent powers and purposes, then that odd neglected corner is a form of prayer, a suspension of design and manipulation and a declaration of faith in the unknown.

The revaluation of weeds, pests and vermin, of fox, jay and cornflower alike, provided a new set of wilderness symbols. They spoke not of luxuriant blooms in distant Edens – the vision of globe-trotting Victorians like the painter Marianne North – but of what was, once at least, small, common and local. Planting them was the human equivalent of rooting. And providing homes for them was neither an attempt to tame them nor, simply, to preserve them as utterly wild beings. It was, fundamentally, an experiment in companionability.

Novel philosophical conjunctions were thus explored. How does one describe a creature that is neither wild nor tame? What is the essence of a relationship in which kinship is felt, and closeness desired, and yet distance and difference are insisted on? This is what the wildlife gardener seeks and yet our words, like our landscapes, tend, in describing it, to imprison us in the phobias of the past. Perhaps the best guide is the early nature poetry of Ted Hughes. Hughes's creatures – pike, crow, thrush – are common, often to the point of drabness, yet on closer inspection embody some indefinable element of mystery, of otherness.

Such minute inspection – one might almost call it introspection, if it did not have a recognizably outer object – may involve penalties as well as benefits. To hear the grass grow and the squirrel's heart beat, George Eliot remarked, is to hear the 'roar which is the other side of silence'. The knowledge of small worlds is often hard to carry in larger ones, although once acquired it seems to persist.

Such at least was the experience of that unusual man of letters and science Loren Eiseley, who sought the common mystery through scufflings in autumn leaves and 'attempts to question beetles in decaying bark'. He did not, of course, find it but he did elucidate with great clarity and beauty a further experimental category of experience, perhaps the most important of those described so far.

Let us call it 'awareness-with-forgetfulness' or 'abandonment-with-recall' and view it as the latest twist in the old theme that man's heart is in nature but his head is his own.

In his wanderings through rural America Eiseley *sees* Kansas wheatfields but *hears* the surf on Cretaceous beaches. In a fin or reptile's foot is an unrealized part of himself. And bathing in the Platte river between the Rocky Mountains and the Gulf of Mexico, he *becomes* the river, 'sliding down the vast tilted face of the continent', yet remembers all that the river cannot – the ancient sea beds and extinct reptiles, the evolutionary concentration of water into life-forms, of which he is one. Consciousness, which in its highest forms is embodied by the city and from which men flee into wilderness, enriches that wilderness and in so doing redeems itself. Heart and head, yet another of those ancient persecuting antinomies, are reconciled.

The proper inhabitant of such small worlds, where a spider's web is the hub of the universe and a few blades of grass form a forest, is the minibeast, that creature which today throngs the literature of environmental education. The minibeast is an extraordinary literary invention, part pet, part chimaera. He is black, scaly and bulbous, unlovable for a generation bred on Pooh, Tigger and mammalian exclusivity, unlovable sometimes to the point of phobia, that ultimate failure of contact. With his relation, the earth animal, he comes in various forms, from the earwig, earthworm and woodlouse to the millipede and the mole. Children, however, appear to be relatively unencumbered with prejudices: teachers talk of the 'cor' or 'coo-er' stage which accompanies the upturning of a rock on the seashore. The primary reaction, in other words, is wonder rather than revulsion, 'coo' not 'ugh'.

Environmental education attempts to build on that sense of wonder, to transform it into connection rather than allowing it to lapse into carelessness. As practised in schools and local authorities, its underlying theme is care for surroundings, the prevention of vandalism in the widest sense of the word. The question then arises: whose surroundings? Who owns and controls them? Who benefits from their protection?

The issue is impossible to discuss without some reference to the debate that crystallized, largely after 1960, around the boundary between school and society. It has posed several questions. Should education take place solely within the school, or in the streets and fields surrounding it? Should learning be active or passive, 'work' or 'play', from books or from life? The debate provided mnemonics like the 'exploding school' or Ivan Illich's 'deschooling society'. It also produced the community college and community education, the former originating in Cambridgeshire in the 1920s, and the adventure playground movement associated with Lady Allen of Hurtwood.

A common theme is the critique of the school, like the modern city and the factory, as a highly specialized institution created largely by industrial society. In it professional teachers deliver a commodity known as education to 'professional' pupils. Only a small piece of the 'whole' individual, the argument runs, is expressed.

In much the same way the industrial city, tightly circumscribed, seals out the country, condemning its residents to diminished lives in which the city is work, the country is play, and suburbia an attempt at a compromise. Against this tradition, in which specialist institutions create specialist people, is set an alternative which canvasses the virtues of personal wholeness, and, more generally, a holistic perspective on life. In human terms this means feeling *more* human than if one were a collection of mechanically efficient parts: it implies some principle of organization which is beyond reductionist analysis. In social terms it means seeing how things link up rather than how they break down. In this respect, its perspective is ecological.

The minibeast is the creature of the emergent heterodoxy. It is the invention of those who believe that 'care for the environment' can be produced only by making the environment local, friendly, familiar, by giving it a form which can be touched and smelled, by speaking of the rainforest only after studying a blade of grass. But the minibeast does not parade on demand like a zoo animal. Its proper habitat is the world of twilight, shadow and mystery spoken of by Lady Allen. Its pursuit must be an adventure and an experiment, the qualities which, as she observed, unite adult research with children's play.

Lady Allen's response in 1960 was to a deprivation perceived as spiritual: the crushing of children's 'vital creative drive' by sterile and arrogant surroundings. Over the same period field study centres were being established in response to widespread intellectual deprivation among young people, and to their alarming ignorance of natural processes. The first centres were set up in the 1930s; there are now over 2,000.

The field study centres, however, were chiefly in the countryside.

Unlike the adventure playground, they were not where children lived. They remained the stuff of day trips, the educational equivalent of tourism. The new heterodoxy sought to combine both needs, spiritual and intellectual, in a way that would do justice to a view of nature which stressed immediacy, personal as well as global.

In 1972 pick axes, compressors and a bulldozer ripped up and reshaped a one-and-a-half-acre black asphalt school playground in Washington, near the centre of Berkeley, California, and a particular mould was broken. Where the asphalt had been, seventy feet from a major arterial road carrying 20,000 vehicles a day through the city, a small green landscape arose. In it were trees, small hills, streams, ponds, bridges, beaches, a wild garden, a community garden for vegetables and flowers, a mosaic of habitats for minibeasts. By the end of the decade 135 species of plants grew there, compared with six beforehand, and forty species of bird were recorded. Equally significant for those responsible, it was 'self-evidently a special place' with a genuine feeling of wilderness.

For the makers of the Washington environmental yard, the old school playground had, like the industrial city, become a metaphor for an obsolete and authoritarian culture. For Robin Moore, the British landscape designer who was one of the yard's initiators, the asphalt was classic 'dumb space', a judgment oddly reminiscent of Julius Nyerere's categorization of cement as 'European soil'. But the new yard was much more than a garden. It was, said Moore, a network of individuals, a community participation process, a new local organization – Friends of the Yard. It was a place for picnics, parties, festivals, for concerts and theatre. It was an educational resource and also a 'kids' HQ', where 'every child can find something stimulating to do day after day . . . under their own volition'. Far from being a liability, it became an asset to the school. When declining rolls threatened the school's existence it was the yard that was cited as a reason why the school should not be closed. Moore coined an eye-catching title for these various roles. He called the yard an 'anarchy zone'.

The use of nature as an outdoor classroom is not new: early reformers like Pestalozzi and Froebel had comparable ideas. But for the large-scale movement which continues to transform the 'old school yard' out of recognition, the Washington environmental yard was probably the chief prototype. The movement began in 1976 with a pilot project to introduce nature reserves into schools in South Yorkshire. The minimal aim was a 'green patch' coupled with a 'wild

pile' of logs and rubble. The organizers were Watch, the junior arm
of the county naturalists' trusts, and an enlightened local authority,
the South Yorkshire county council. By 1980 there were almost
ninety school reserves around Sheffield and Barnsley and the
example was spreading to scores of other towns. In 1983 alone the
British Trust for Conservation Volunteers, which has played a leading
role in the school greening movement, worked with 20,000 children
in over 500 schools and was aiming to cover 5,000 schools by 1990.
Many more initiatives have come from urban wildlife groups, nature
trusts and forward-looking local education authorities like Avon and
Cheshire.

But to lump all these projects together as 'school nature reserves'
is to miss not only the variety of landscapes created but the sense of
challenge and excitement of those taking part, the crumbling of
specialisms and boundaries, and above all the responsibility
shouldered by children in the shaping of land – their own land.
Insect gardens, for spiders, butterflies and other minibeasts, have
become almost commonplace. So too have wild corners – bramble
patches, piles of leaves, rotting wood. Children have gone on to

Plate 9.1   Creating a school garden in Hull

build walls, rockeries, paths and seating. They have laid out herb gardens, bog gardens, heather gardens, bird gardens with hides, vegetable gardens, allotments. They have planted and managed tree nurseries, hedgerows, thickets, small woodlands, orchards. In Gateshead they built a duck pond, in Leeds and Cardiff a fully-fledged natural park, in Birmingham a small farm.

All this has happened *within* school grounds, usually on old tarmac, lawns or playing fields. But shaping land is evidently addictive. What began as an experiment has burst out of the test-tube. As the small worlds grew larger and school boundaries were bypassed, children's landscapes became not so much laboratories as living-grounds – 'real' places for grown-up people. At the end of 1981 the Newcastle and Gateshead schools and community project was launched and within two years thirty schools were managing their own miniaturized safari parks. But outside the school children also helped to build three large and ambitious wildlife and nature parks, at Windy Nook in Gateshead and Walkergate and Benwell in Newcastle.

Five rows of old terraced houses, overlooked by a cluster of factory chimneys, once stood at Benwell. They were demolished in the early 1970s and a decade later only patches of cobbled street remained amidst the rubble and scrubby grass. In the summer of 1982 earth-moving began, a classroom was set up, a pond dug, and 400 children planted trees. The next year 600 children took part in a 'grass-in', planting seed. With Windy Nook and Benwell together children from fifty schools were involved.

Benwell is a tough area and both residents and children were initially sceptical. One group of 14- and 15-year-old girls, said one observer, came 'determined not to enjoy it' – but they did. After a ceremonial release of frogs in March 1983 children were heard issuing warnings against throwing stones in the pond in case they 'hit a frog on the head'. Jim Grant, a Benwell teacher who filmed the old houses being demolished, believes the children's intimate contact with the park is vital. 'They've seen it all growing,' he says, 'and as time goes on I think they'll feel it's more and more theirs.'

The Newcastle project is one of many. In Edinburgh the Environmental Resource Centre has since 1976 done outstanding work in fostering the message that community development and environmental conservation go hand in hand. The centre's philosophy is that self-help, an individual's power to produce visible change in his surroundings, makes for whole people and that this is the way to

Plate 9.2 Children, minibeasts and earth animals: a close encounter on the home ground – St Jude's nature garden, Bethnal Green, east London

conquer apathy, ignorance and vandalism. For Graham White, one of its originators, planting a tree – merely a pleasant act for an adult – may, for a child, have lifelong emotional significance. The greatest threat to the natural heritage, according to White, is the 'apathy of the vast mass of people living in the increasingly harsh environments of our inner cities'.

With the centre's help a school – Dumbryden Primary, in the vast high-rise housing estate of Wester Hailes – made its own allotment and orchard, producing £100 worth of crops at its first harvest, contributing to school dinners and raising money from harvest sales for seeds and tools. Schoolchildren and their parents joined in community planting projects in Pilton on new landscapes designed by consultants working for the Scottish Development Agency; a marked lack of vandalism was one notable result. Unemployed fathers built a flower garden and vegetable patch for their children at Craigmuir primary and nursery schools. Schoolchildren at Easthouses took part in a rescue of a wood near their school which was scarred by dumping and vandalism. Young volunteers, all without jobs, have made a 'mother and toddler' garden in a community centre, built nesting boxes and bird tables for old people's homes and worked widely with schools.

The centre has also played a leading role in introducing the ideas of the American naturalist and educator Joseph Cornell to school-children. Cornell's fascinating book *Sharing Nature with Children*, enormously influential in the United States and published in Britain in 1981, contains dozens of 'games' which by way of direct sensory experience teach ecological concepts like the functioning of food chains. Learning comes alive as personal drama. In 'Meet a Tree', for example, the child is blindfolded, asked to feel a tree – the shape and texture of bark and branch, the smell of blossom, the taste of leaf, the sound of insects – and is then led away and, with the blindfold removed, is challenged to find that particular tree. According to Cornell, they almost always succeed. Observers describe the effect on children as astonishing.

Schoolchildren in north London have experienced an even more dramatic version of encounter learning in the 'acclimatization workshops' pioneered in the 1960s by the American Steve van Matre. Rooted in the concepts of Piaget, acclimatization or 'earth education' has children walking barefoot in a bog, crawling in tall grass, travelling through the treetops in a bosun's chair. It teaches 'earth magic' through drama, ceremonial, night-time exploration, detailed examination of 'micro-parks', the creation of 'leaf graveyards'. Its effective British headquarters is Capel Manor field studies centre in Enfield where Stewart Anthony, the field studies officer, has worked to disseminate van Matre's ideas in Britain. The impact of acclimatization on children at Capel Manor has been dramatic.

All these developments, from the school nature reserves movement to acclimatization, are primers for rediscovering lost connections. Their common demand is for nature to be accessible, a present reality, not an occasional dream. Walking barefoot in a bog requires a real bog, cold, wet and messy, in the here and now. Textbook and televised bogs will not do, nor will a bog seen by appointment at six-monthly intervals. And since so many 'country' experiences, once so

Plate 9.3   Benwell nature park, Newcastle upon Tyne: learning, idling and fun

commonly available, have been lost to generations of urban children, they can hardly be blamed for their ignorance. A survey in 1981 showed that 82 per cent of 7-year-old children in south London did not know that peas grew in pods. Many children who visit Capel Manor are simply unaware that milk, eggs or meat come from animals.

Ignorance on this scale promotes conspicuous consumption and extravagant waste. It is the mark of a city cut off from the countryside, so that crossing the boundary induces a kind of culture shock. 'The silence, the space, the solitude so valued by many . . . may be actually threatening to a youngster impelled into it from the heart of a city.' That is the conclusion of Rob Collister, a Snowdon outdoor pursuits instructor who has begun to doubt the value of bringing untutored children into the mountains. The city child, says Collister, 'reacts by yelling raucously into an empty cwm, shattering with stones the mirror of an upland lake, and throwing aside his crisp packet and empty pop-can.' A Snowdonia in which tin cans and broken glass glitter from every lake bed, in which every summit reveals a circle of dried milk sachets and polythene bags, is thus the penalty of using the mountains as an 'outdoor gymnasium'.

Yet this is precisely the approach fostered by the old school playground and the cast of mind that goes with it. Iona and Peter Opie, in their intriguing collection *Children's Games in Street and Playground*, found playground behaviour 'markedly more aggressive' than games in the street or in wild places. Playgrounds are also dangerous places physically: the impact of a fall on tarmac is 150 times greater than on a soft surface like sand. They are another version of 'absolute space', the term used by economists to describe land in cities. Because they are devoid of feature, topography and character they do not feed children's minds. Playground designer Lin Simonon likens them to an army parade ground. The fault, one speaker told a conference on schoolyard landscapes in 1981, lay with the headmasters. 'Many see themselves as chief warders,' he added acidly. 'They would have wall to wall tarmac and guards round the perimeter.'

Such an authoritarian response views vandalism as a failure of individual reason or social discipline rather than a loss of connection between an individual and his surroundings. Environmental education thus becomes a subtler version of that time-honoured injunction: Keep Out.

The insights of environmental psychology and ethology point to an intimate connection between vandalism and notions of territoriality, ownership and control. The paint-sprayed wall and the uprooted tree is thus a muscling-in on the business of landscape design, a point Jamie McCullough makes brutally clear in his account of Meanwhile Gardens. Hilary Peters at Surrey Docks city farm in London was plagued by vandals who felt left out. 'The only way of coping with them,' she said, 'is to let them in.' Vandalism, in effect, is a kind of land-grab. 'Keep Out' has lost much of its power to subdue.

Ways of seeing and behaving merge. At William Curtis Ecological Park schoolchildren studied poetry and compost heaps. At Dumbryden primary on their allotment they study art and home economics. Far more than their parents, as mobility studies have shown, children are trapped in the city; their need for a fine-grained, home landscape of their own is so much the greater. In the small worlds of garden, school reserve and nature park, the holly leaf-miner is met and the language of webs, cycles and linkages is learned, intimately, at close range and first hand.

In 1981 a distinguished geneticist and a dozen colleagues attempted to formulate a new code of environmental ethics. It was a task of global import, addressed as Britain's contribution to the 1980

World Conservation Strategy. By general agreement it failed – chiefly because it was attempting a virtually impossible task. 'No important change in ethics,' wrote Aldo Leopold, 'was ever accomplished without an internal change in our intellectual emphasis, loyalties, affections and convictions.' Ethics, in other words, are changed by events and emotions, not by edict. A young Swansea valley tree-planter, interviewed over five years later about what it had meant to help reclaim a scarred landscape, touched the point more nearly. 'Adults didn't have wildlife when they were younger,' he replied, 'so they didn't learn to care for it.'

Environmental education ultimately ceases to have anything to do with schools. In 1982, while the committee of thirteen were constructing their code of ethics, the first Earthways workshop was held at Lake Geneva, Wisconsin. Developed as part of the acclimatization approach, this centres on young people 'organized as a clan to champion the existence of one of the earth's natural communities' – in effect, a kind of totemism. Each clan will 'seek the end of the rainbow' by following seven paths, one for each colour of the spectrum. These focus on a special way of interacting with the natural world: perceiving, exploring, understanding, feeling, silencing, harmonizing, simplifying. Individuals then embark on a personal odyssey to 'solidify . . . their new relationship with the earth and its life'. They may walk across a continent, accompany a river from its source to the sea. The object is a special relationship with their 'real home', the planet earth.

In the upper Mersey valley in Manchester, adult residents are undergoing their own kind of acclimatization – in this case, to a green riverside landscape slowly being rescued from almost two centuries of tipping, poisoning and pollution. Walks are organized, at dawn to hear the bird chorus, at midnight to see the full moon. There is also an 'earth magic' walk, described as a 'sensuous immersion in the natural world'. What in small worlds is education may in larger worlds, it seems, amount to transformation.

10

EARTHWORKS

'Your future, folks, is urban hippies, and a good job too.'

(Bill Mollison, address to Soil Association of South Australia, 24 February 1983)

Imagine Los Angeles gradually transformed into a giant orchard laced with bikepaths and a solar freight rail, tended by thousands of solar co-op communities.

(From Los Angeles, A History of the Future, *Citizen Planners, Venice, California, 1984)*

In the parched and broiling summer of 1976, a score of unemployed teenagers laid siege to a rubble-strewn and bramble-infested acre of land next to a steep railway embankment in the St Werburgh's district of Bristol, half a mile from the city centre. Starting work at 6 a.m., to avoid the worst of the heat, the teenagers dug up half a ton of roots from iron-hard ground, carted away 5,000 bottles, hacked down enough scrub, grass and bramble to produce ten tons of compost, drew off 600 gallons of water from a small stream nearby, and put up a strained-wire fence. Finally the land was ready to produce food: there were twenty-one plots in all, occupied by members of the newly formed local allotments association. The job had taken three months.

The back-breaking work at St Werburgh's marked the start of the Bristol city land-use project, an initiative which sought to apply traditional rural skills of craftsmanship and land husbandry to the social problems of the cities. Yet it was as much a keen awareness of unused assets that prompted the project: the growing army of young unemployed; the tons of waste paper, sewage and spent hops; the timber of dead and diseased elm trees serving little purpose but to disfigure the skyline.

Across the city centre, meanwhile, that same hot summer saw the stirrings of an even bolder venture: a fully-fledged farm. The prospective site – five acres of flattened slums 100 yards from the Paddington-to-Cornwall railway line – appeared unpromising. The land was piled high with scrap and junk. Gypsies camped there. Bristol council, worried about health risks, was somewhat despairingly proposing a lorry park.

The first sign of a change on those five acres at Bedminster, below the Inter-City railway line along Windmill Hill, was a crop of bright

Plate 10.1   Farmyard at Windmill Hill city farm, Bristol

yellow mustard, planted as much in self-advertisement as to fertilize impoverished land. By 1978 the farmers had moved in. The rubble was used to sculpt land and fill drystone walls. Family gardening plots were laid out, alongside paddocks, hedges and fruit trees. Perhaps most important, a proper farmyard was built, with a duck pond, stables and outbuildings. The pond was old-fashioned, 'puddled' with clay from Bristol docks. In the small fields and outbuildings, animals – sheep, goats, pigs, chickens – were bred, farm products sold. So too was fowl at Christmas. Eggs, yogurt and cheese became staples. Bacon and ham were cured, with pork especially popular. Piglets continue to fetch good market prices as weaners.

The examples set at St Werburgh's and Bedminster helped to produce some remarkable multiplier effects. Within five years it was possible to see farming, forestry and haymaking inside the Bristol city boundaries, to see old techniques of husbandry – the laid hedge, the pioneer fence, even the Devon bank – restoring damaged land. Two more city or community farms, at St Werburgh's and Hartcliffe, a giant urban-fringe housing estate built to accommodate the refugees

Plate 10.2   Fields at Windmill Hill city farm, Bristol

from clearances of slums like Bedminster, followed Windmill Hill
farm.

At Stockwood, on the southern outskirts of Bristol, seventy acres of
'trapped' fringe farmland designated as council open space were
threatened with vandalism and 'horseyculture'. The Avon Wildlife
Trust proposed a radical new use to the council: combining the
restoration of old farming practices with ecological management
aimed at enriching its wildlife. Above all, said the trust, local people
should be brought in, not only to help in pond clearance and land-
work but to take over its management, perhaps acting as wardens or
'adopting' a copse or a hedgerow. To the trust's surprise, the council
agreed and handed the land over.

At Brandon Hill, meanwhile, the city-centre park which contains
one of Bristol's best-known landmarks, the Cabot tower, 1981 saw
the first summer hay crop. The aim was to reduce maintenance costs
and also to cut down on nutrient levels so that wildflowers would
flourish. The hay crop went to St Werburgh's city farm. The same
year saw Bristol council apply to the Forestry Commission for grant

aid under a Basis III dedication covering 630 acres of its woodland. This was no ordinary commercial forestry, however, since the policy laid a heavy emphasis on ecology, amenity and aesthetics – the intangibles of landscape beauty. It was a new approach in a new setting: the urban forest.

For the discerning visitor, however, Bristol had even more to show. In a ramshackle old row of terraced dwellings overlooking the blackberry and apple plots of Windmill Hill farm is a house dedicated to the sun. On the wall of an upper floor there is a large, stylized depiction of it, part icon, part isometric – a segmented yellow orb inscribed runically with a poem by Robert Frost, a proverb of the Nootka Indians, some translucent Japanese verse. But it is sun-worship of a highly cerebral variety. The technology of solar magic – magic, that is, in its generic modern sense, meaning the manipulation of an ultimately mystical force – consists of trombe walls, solar panels, triple-glazed windows, shredded-paper loft insulation, heat pumps, insulated cooker and haybox, underfloor polystyrene, and even a small windmill.

The 'future city home' is part of the Urban Centre for Appropriate Technology, developed since 1979 by a group of planners, gardeners, engineers, architects and teachers. Designed as an urban version of the outstandingly successful Centre for Alternative Technology, Machynlleth in North Wales, it is much more than an energy-saving house. It is a community energy workshop where schoolchildren and adults can learn do-it-yourself conservation, including skills like secondary glazing and solar panelling. It is an energy information centre and a draught-proofing and insulation service: 'granny-lagging' in the colloquial. It is a fruit, herb and vegetable garden practising intensive but organic backyard cultivation and crop-rotation methods: growing on raised beds and trellises, in tubs, greenhouses and cold-frames. It is a smallholding, rearing hens and rabbits, keeping bees but also preserving a wild area with a pond. It is a 'conservation kitchen' where wastes are separated, recycled, and composted and where domestic thrift extends to a compost toilet and a digester to produce biogas from human and animal waste. The future city home is both a glimpse of a possible urban future and an experiment with an unavoidable urban present. Its gardeners test mixtures of local and clay soil, compost and manure to see which most successfully cuts down lead and cadmium levels. Noise and vibrational power is collected from rush-hour traffic and transformed experimentally into electricity.

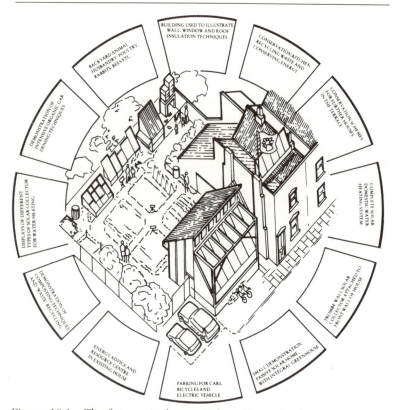

Figure 10.1 The future city home: Urban Centre for Appropriate
Technology isometric

The future city home began to take shape at the start of 1982. In
Bedminster, meanwhile, recycling has assumed new and impressive
dimensions. In a former rubber factory, now leased from a
manufacturer of hot air balloons, Resourcesaver, begun in 1980, has
created 100 jobs out of city wastes. Launched by Avon Friends of the
Earth, the '3 Rs' centre – repair, re-use and recycling – has reversed
the consumer product cycle, finding raw materials in rubbish,
displaying an ingenuity which bewilders those taught to view the
dustbin as a form of domestic black hole. Waste paper, rags, oil,
glass, plastics and aluminium are collected from homes and offices
by horse and cart. There is a scrapstore and a linked craft and toy
workshop, turning scrounged or discarded junk into school and
playgroup materials. In a repairs unit, furniture and consumer goods
are renovated. Small businesses have been set up: potting,

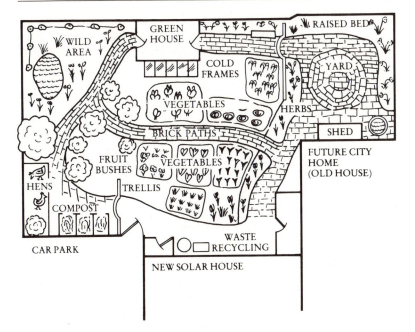

Figure 10.2   The garden of the future city home

woodworking, bicycle pannier manufacture, Treesaver recycled paper products, computer sciences, printing. Car-sharing and energy-saving schemes are run. Collections cover Bristol and Bath; householders themselves undertake much of the separation of materials and the bundling up.

What unites these ventures in Bristol is the attempt to blend pre-industrial features into the emerging post-industrial city. The land-use project and the work at Hartcliffe community park farm owe much to the Green Cure Trust set up by Bob Lorraine, a retired engineer, and David Gordon, a former Devon farmer. City land, says the trust, must be reinstated as a 'living organic thing, a basic and complex element in the life support system'. It is a vital neighbourhood resource. Working it can foster 'originality, enterprise, stability and responsibility'. Gordon believes that city farming is a reversal of the eighteenth-century enclosures that drove people into the cities. He calls it 'taking back the commons'.

The emergence of the land ethic inside urban boundaries looks back to the pre-industrial city, before food production was squeezed out by a combination of green belts, infill, population pressures and *politesse*. It is an attempt to integrate economic and social goals, to find a technology that does not end in waste and damage. It emphasizes care, craftsmanship, husbandry and heritage, the local and the low-cost. It reinstates the primary production of food, fuel, timber and ore in the tertiary and quarternary zones of retailing, services and banking. It disrupts industrial land-use patterns by opening up a rural frontier in the heart of the town. It also disrupts consumption patterns, by revealing the potential of dustbin, sewer and drain. It even calls into question that most ancient and parasitic role of cities, the manipulation of rural surplus product. The city, it says, should generate its own surplus.

For the new breed of urban earthworker, the city in its full mid-twentieth-century florescence represented much of what was wrong with a civilization desperately out of sympathy with its global hinterland. Earthworking strikes at the headquarters of this civiliz-ation, employing ecological cycles/thinking to demonstrate that neither people nor products need undergo some drastic and degenerative change when they move from country to city. In these ways, it looks forward to a new post-industrial city.

City farms began, in Britain, in protest – a squat in disused stables, sheds and allotments above a railway embankment in Kentish Town, north London, in 1971. The squat was led by Ed Berman, the grass-roots activist who a decade later was to be found advising the Government on urban regeneration. Thus was born City Farm I. From the late 1970s the growth in city farming has been dramatic. In 1976, when the city farms advisory service was launched under the tutelage of the Inter Action Trust, there was a handful of projects, Surrey Docks, Kentish Town and Freightliners, all in London, among the best-known. In 1980 the National Federation of City Farms was launched with thirty-three member farms. By 1984 the number had doubled – sixty-five were working or shortly to become so – and protest had been subsumed in wider objectives of agricultural experimentation and community building.

In practice this meant that a city farm was not only a park, garden, allotment and nature reserve. It could also be a school, a workshop, a playground and a social club. In some cases it was the focus for an entire neighbourhood. A sunny morning in the school holidays at Windmill Hill in Bristol, for example, sees a remarkable bustle and chatter round the duck pond. Here are gardeners from the family plots, or teenagers learning about computers and photography or youngsters playing on the hard sports pitch. Here are parents bringing their children to the crèche or to the adventure garden or to the rumpus room or to the pre-school play centre. Here, alongside the farmshop and café, are clubs for the socially isolated, horticultural therapy sessions for the mentally ill, craft instruction for the single parent. A powerful impression is created of a strange hybrid organization in which work, pleasure, education and healing – functions usually kept rigidly separate from one another – co-exist and intermingle. The atmosphere reflects this, a mixture of country market, village fête and small-town high street. If such a thing is a farm, one wonders what, or who, is being cultivated.

City farms are attempts to bandage up the wounds of inner-urban dereliction, the scarred landscapes and hurt people. They are also experiments: in urban productivity and resource reclamation, in a 'new' agriculture, kinder to the earth, to animals and to people. Leftovers from take-away shops, greengrocers, breweries and food factories provide animal feed or compost. Other people's cast-outs are scrounged, renovated and resurrected. Chicken houses are made from BBC scenery, ponds from two-level baths, hills, banks and mounds from car tyres. City farms have harvested carp for Chinese restaurants and worms for anglers; they have produced mushrooms, artichokes, nettles, dandelions, 'ethnic' vegetables like fenugreek, used in curries, gourds, melons and jute. Crops have been grown on roofs and in water, hydroponically. City farmers attempt to construct richer, more rounded and enduring plant-animal relationships. Fruit trees, for example, are grown for amenity and ornamental shrubs for fodder and nitrogen-fixing. Duck ponds are kept clean by colonies of freshwater mussels. Often this means older, tougher and rarer breeds of animal better equipped for urban foraging, like the Soay sheep or the Gloucester Old Spot pig. In such cases species conservation also plays a part. Or it might involve the 'chicken tractor', a series of runs in which the hens live free-range because they dig, harvest and forage for themselves.

City farming, however, was merely the most visible of a wide range of earthworking initiatives in the cities, from bottle banks and neighbourhood energy schemes to municipal recycling projects, from bee-keeping and timber growing to co-operative food pro-duction. Such initiatives, often, have been local, spontaneous, unco-ordinated, and therefore unnoticed. In Edinburgh, for example, flat-dwellers produced food from pensioners' gardens, grown too large for their owners to manage. In Sparkbrook, Birmingham, weed-choked gardens have been joined together to form Ashram Acres, an exercise in small-scale land reclamation, animal husbandry and horticulture undertaken by local Asian residents, many of them small farmers by origin. In St Helens, between Manchester and Liverpool, a father-and-son team has created a rainbow trout fishery out of a derelict claypit bordered by a colliery waste heap, a power station and a steelworks.

Figure 10.3   Ashram Acres: the ideal

Some groups, like Liverpool's aptly named Diggers, have adopted an active role campaigning for the productive community use of city wasteland. Others, grown tired of waiting, have simply squatted. Prospect Walk in Hackney, east London, where once two rows of prefabs stood, was the scene of one such take-over. The old pavement, complete with lamp-post, became a path between allotments.

Even less visible are the many community recycling ventures set up in the wake of recession and unemployment, most of them tucked away anonymously behind some little-altered Victorian factory or warehouse facade. Bristol's Resourcesaver has counterparts in dozens of other towns and cities, each a small centre of economic

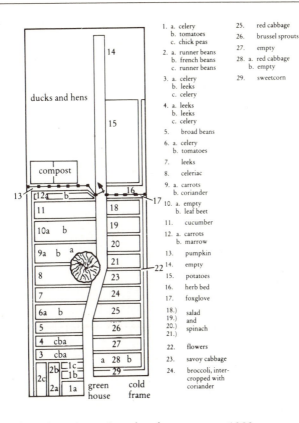

1. a. celery
   b. tomatoes
   c. chick peas
2. a. runner beans
   b. french beans
   c. runner beans
3. a. celery
   b. leeks
   c. celery
4. a. leeks
   b. leeks
   c. celery
5. broad beans
6. a. celery
   b. tomatoes
7. leeks
8. celeriac
9. a. carrots
   b. coriander
10. a. empty
    b. leaf beet
11. cucumber
12. a. carrots
    b. marrow
13. pumpkin
14. empty
15. potatoes
16. herb bed
17. foxglove
18.)
19.) salad
20.) and
21.) spinach
22. flowers
23. savoy cabbage
24. broccoli, inter-
    cropped with
    coriander
25. red cabbage
26. brussel sprouts
27. empty
28. a. red cabbage
    b. empty
29. sweetcorn

Figure 10.4 Ashram's garden plan, late summer 1983

subversion expanding the concept of product *life* into materials and energy *cycle*. They are called, variously, recycling centres, scrap stores, or wastesavers. Edinburgh has seen a community thrift shop, Leeds the ambitiously titled Save Waste and Prosper project. In Walsall they called themselves the Wombles. Hackney's Brass Tacks Workshop, among the first to demonstrate how recycling could create broader job-generating spin-offs, was the forerunner of a national network of mutual aid centres. The ill-fated Greater London Council itself sponsored a network of thirty-six recycling centres.

A small but growing number of far-sighted local authorities have meanwhile braved the market aberrations that are the bane of recyclers and experimented on a larger scale. In 1985 the bottle bank concept, extended to cans via the Save-a-Can scheme launched in

Leeds in 1982, was elaborated further with the first 'reverse vending machine', set up in Northampton; the machine pays a penny for two empty cans. Reclaiming household rubbish, or extracting fuel or electricity from it, has been attempted with varying degrees of success; in 1985 an estimated ten plants were operating waste-to-energy schemes. An early experiment in municipal composting at Worthing has been followed up in Glasgow, where grass cuttings from parks produced biogas for heating greenhouses, and Bristol, where fermenting sewage sludge supplies enough electricity to the national grid for a village of 2,000 people. A district heating system for houses at Byker in Newcastle is fuelled by urban waste. Computerized waste profiles, drawn up in Birmingham as part of a waste-to-fuel project in 1984, have highlighted the link between land-use, living density and resource conservation. Inner-city dustbins, for example, generated by far the most vegetable waste.

Statistical indices can chart only the barest outlines of these developments. They can tell us, for example, that by 1985 there were 2,000 bottle banks in 740 towns and cities, that 100 voluntary neighbourhood energy schemes were set up between 1981 and 1985, that the waiting list for allotments increased enormously in the 1970s, from 10,000 in 1969 to 120,000 in 1977. But for a true portrait of the earthworker and his attitudes we have to rely on the sort of individual examples mentioned above, and what these speak of is an urban society slowly rediscovering its lines of supply. Behind the earthworks, ultimately, lies a vision of the city not only as a source of food and fuel but as a place where new, productive and sustainable relationships with the earth, its products and cycles, are coaxed into life and vigour, not suffocated at birth.

Cities in the developed world generally, notably the United States, present a similar picture. By 1981, for example, there were an estimated two million community gardeners in the US, tending plots in public parks, churchyards, schoolgrounds, below power-lines and on the roofs of hospitals and universities. They have been supported by a federal urban gardening programme and a thriving network of voluntary greening groups. Results have been striking. In 1977 the first corn grew on empty lots in New York. By 1985 local groups forming the Neighbourhood Open Space Coalition had created 448 community gardens and parks on 150 acres of vacant land in New York City alone. In 1978 Newark, New Jersey, once the scene of race riots, celebrated its first annual harvest banquet with local produce. In the nearby Bronx the appropriately named Frontier Development

Corporation advises residents on nutrition and horticulture and markets Zoodoo, a mixture of composted vegetable waste and dung from Bronx Zoo.

By 1982 more than 200 American cities were estimated to be recycling waste. Many of the programmes were built on voluntary ventures which had proved more resilient than the overambitious municipal and federally funded schemes. In St Paul, Minnesota, more fundamental changes were evident. In 1969 two grass-roots recyclers, Joanne and Carroll Nelson, started a backyard collection centre they called Recycling Unlimited. By 1984 it had grown into a non-profit-making business with a planned annual capacity of 18,000 tons. The city itself had meanwhile embarked upon a 'home-grown economy' programme, aiming to make St Paul the first self-reliant city in the United States. Recycling Unlimited was its model of a local industry using local materials to generate local jobs.

The philosophical and technical support for the St Paul Project, and many similar ventures, has been provided by the Washington-based Institute for Local Self-Reliance. Founded in 1974, the Institute has a distinctive vision of the city. It sees it as a nation-state operating a balance of payments with the outside world and grown 'dangerously dependent' – importing too much, over-reliant on absentee landlords and factory-owners, wasting its productive potential, both human and technical. It argues that recovering the 'gold in garbage' effectively moves the farm, wells and mines into the cities. The sewage plant thus becomes a fertilizer factory, the basement boiler a power plant. It canvasses sewerless cities and home energy production.

But practical earthworking is inseparable, in this view, from political and social development: it demands from the individual a new awareness of his powers and responsibilities. In the new city-state the citizens, no longer mere voters and consumers, are 'producers of wealth and managers of the city's future'. Earthworking, in other words, is a practical, intimate and unhistoric way of changing the world.

The social and political dimensions of earthworking technologies are thus as fundamental as their ecological aspects. Green thought, from this perspective, fuses with a more recognizable political radicalism which opposes the centralization and concentration of power not merely because it is politically illiberal – this is the orthodox nineteenth-century stance – but because it is socially inefficient. It thus has much in common with both the new conservatism and with an older rural and democratic, indeed anarchic, socialism – the small-scale mutualism, associated with figures like Proudhon, Tolstoy, Morris and Kropotkin, which was subsequently overwhelmed by the rise of the urban-dominated Marxist collective.

But earthworking, as the examples quoted above suggest, is more than a sentimental return to the land. It is an attempt to marry two deeply divergent intellectual and technological traditions: on the one hand, man the creator, inventor and artificer, clustering together in purpose-built think-tanks and specialist test-laboratories, the cities; on the other hand, man the creature, the worker by heart and sinew, more isolated and vulnerable but tuned to the profounder harmonies increasingly identified with the countryside and with rural crafts and food production.

These themes emerge perhaps most strongly in permaculture, the 'perennial agriculture for human settlements' devised by the Australian Bill Mollison. Mollison is an extraordinary figure. Living alone as a trapper, logger and farmer in the Australian bush for thirteen years until he was 28, he founded an organic gardening society in 1972 because of a self-confessedly naive belief in 'clean food', discovered distant threats to his garden soil – pesticides, acid rain, lead in petrol – and slowly elaborated the philosophy which he articulated in 1978 in *Permaculture One* and *Two*. In 1981 he was awarded the Alternative Nobel Prize. By 1983 there were twenty-eight permaculture groups on Australia, sixteen in the United States and thirty more in a dozen other countries. In Britain its most important impact has been on the city farms movement, through which permaculture concepts are now spreading more generally into urban gardening and greening.

Permaculture is a complex, fascinating and yet remarkably coherent set of ideas. It is a synthesis of philosophy, science and inspired insight, of ancient systems of cultivation and new silicon-based technologies. It is also a challenge to contemporary agriculture and to cities, issued in the name of a new breed of urban peasant. 'There is nowhere in the city that we should grow food: nowhere but the city where we must grow food,' says Mollison. Gardening is a poor word to describe it. Rather it is biological harmonics, the orchestration of plants, people, earth and animals to produce ever-richer connections.

In permaculture, old divisions wither. Permaculturalists build living fences, grow fuel, design 'beescapes', plant fodder hedges. A washing line is a solar drier. Gardens are three-dimensional, climbing up walls and trellises on to roofs and balconies, sheathing buildings in living tissue, so-called 'biotecture'. Goats mow lawns, geese weed potato plots, worms make soil. Computerized 'personnel' profiles on plant and animal species are built up, to see where they thrive best. Key themes are co-operation, connections, peaceful co-existence: individual self-expression within a broader cosmic rhythm.

That, at least, is the ideal. A representative story is that of Gil and Meredith Freeman who stayed behind in the city – Melbourne, Australia – in the early 1970s when many of their friends were abandoning it for the countryside. With three other families they bought four houses, took down the fences and christened their venture Compost. By 1983 the Freemans' own 350-square-metre permaculture garden was producing 80 per cent of the food needed for their family of four. Gil Freeman describes it in terms reminiscent of Jenny Owen's Leicester garden: 'a frantic mixture all gathered in ... and growing healthily together'. Plant disease, he reports, has dropped dramatically, a fact largely attributed to the sheer quantity and diversity of intercropping.

An equally representative story is that of a rooftop group in Camden, north London. Intent on creating a mulch garden, they collected plastic lining sheets, sacks of food waste, straw and horse manure. Local residents, mistaking the kerbside pile for an impromptu refuse tip, added discarded filing cabinets, an oil lamp, an old glass-fibre car bonnet, a flock bed and a horse hair mattress. Undeterred, the permaculturalists incorporated every donated item into their garden: 'more from less' is a favourite theme.

If earthworks demonstrate the productive potential of cities, the concept of the urban forest takes this an important psychological stage further. The urban forest offers fuel, energy, jobs, raw materials for craft industries and import substitution. In 1985, for example, the Inner London borough of Tower Hamlets began the planting of ten derelict sites, two of them at the northern entrance to the Blackwall Tunnel, as part of its forestry management plan, designed to provide wood for local timber companies and craft industries. But equally important, it offers a powerful solvent of the city's identity. The forest, particularly for European cultures, remains what many people mean by wilderness. It signifies the densest, the most 'extreme', form of naturalistic landscaping. Merely to yoke it to the idea of a city shows how earnestly the union of two ancient opposites is desired.

The urban forest may be a large-scale exercise in afforestation on the urban fringe, as in Bos Park, Amsterdam, or West Forest Park in Copenhagen, the latter begun in 1967 and planned to cover 3,200 acres. It may be smaller and nearer a city centre, as in Central Forest Park on 130 acres of derelict land at Stoke. It may be a rich, dense framework for housing, as in Warrington new town, or thickly planted linear belts as at Milton Keynes. A minimum size is often involved. Sweden's urban forests, for example, used largely for timber production, cover at least 125 to 250 acres and must be easily reachable on foot, bicycle, ski or public transport from the city. Physical factors like tree density may be subordinated to mental constructs. In Oakland, a city of 350,000 in California, the urban forest is a collectivity of all the city trees rather than a thick tract of woodland. The forest is regarded as a vital conceptual tool for landscape managers.

Particularly since the 1970s, landscape planners have sought increasingly to *make forests* rather than *plant trees*, a vitally important change of emphasis. Tree-planting, gently ameliorative, gave us the traditional cityscape of the London square. With the large-scale urban reclamation of the 1970s, coupled with losses from Dutch elm disease and other causes, it took on new dimensions. In the decade from 1974, for example, nine million trees were planted in Greater Manchester, an area of mature tree cover at least three times as large as the New Forest. The New Towns programme has also borne witness to this changing emphasis. A new generosity in the provision of green space, of which Harlow and Hemel Hempstead are the best early examples, found particular expression at Cumbernauld where, whatever failings were later to emerge in its vision of urbanity, nearly a thousand acres of woodland were planted. In the 'forest city' of Milton Keynes, however, where the New Towns programme has culminated, wooded road verges alone covered almost twice as great an area. The urban forest, like the wildwood in *The Wind in the Willows*, may thicken imperceptibly until one day we look out and it surrounds us.

The launch of two new voluntary groups, only a decade apart, illustrates another important shift of emphasis. In 1972 the Woodland Trust was founded with the aim of *saving* threatened *rural* woodlands. It has since proved one of Britain's most successful conservation bodies. In 1982 the Town Trees Trust was founded with the aim of *creating* new *urban* woodlands, on the land left vacant by recession. Each Town Trees site, of which there were seven in London within two years, the first at Lismore Circus in Camden, is a tree nursery as well as a small landscaped park. 'Production is visible,' as David Embling, the trust's founder, explains.

Both bodies, however, capture the profound relationships evolving between human beings, plants and landscapes. The Woodland Trust's 'commemorative grove' programme, for example, in which small woods can be dedicated in memory of loved ones, represents a variant of animism with which our pagan Celtic ancestors would have been thoroughly comfortable. The Town Trees Trust involves local people, particularly children, in design and planting; a school or a tenants' association may 'adopt' a wood. The trust also lays great stress on the wider environmental benefits of urban woodlands.

Ecological and climatic studies, for example, have demonstrated what many Victorian landscape designers felt intuitively to be true. Vegetation, particularly trees, helps to purify cities. Woodlands reduce the climatic extremes of cities, moderate wind turbulence and temperatures, counteract aridity by aiding the dew cycle, filter and metabolize pollutants, reduce noise and oxygenate the atmosphere. In Russia 'sanitary clearance zones' of woodland have been established. In Stuttgart, West Germany, green space reaching into the city from surrounding rural hillsides is designed as a conduit for cool air to drain down the slopes at night and flush out pollutants. The Town Trees Trust calls this 'air conditioning' cities. Research in biometeorology suggests that the gains for human health are substantial.

On the model of the two trusts, 'community woodlands', managed, worked and wardened by local groups, are proliferating in cities. Typical is Peasecroft Wood, eight acres of sycamore in Wolverhampton owned by the borough council which grant aids Bilston Conservation Association to run it. Conservation volunteers are trained there in techniques like pollarding and coppicing. Students from local schools and a polytechnic use it for field studies. But neighbourhood involvement can also mean a simple 'adopt a tree' programme, of the sort launched as part of Operation Greenup in Florida in 1979. Tree survival rates in such programmes are often staggering. In Fort Lauderdale, for example, they reached 95 per cent.

One of the best clues to the urban forest of the future, however, comes from Bristol. In 1977 the city initiated its woodland advisory panel, where foresters and conservationists sit alongside local amenity and wildlife representatives. In 1980 its first woodland officer was appointed and in 1981 a forestry grant applied for. The city meanwhile produced a new woodland estate working plan.

Under the management guidelines laid down in this ambitious document, the woodland canopy, the 'outline', is retained and trees are kept beyond economic maturity. Rotation forestry, the routine blitzing of large tracts of woodland, is outlawed. Treatment is of 'small, random irregular' areas, aiming at a multi-storeyed woodland and greater wildlife diversity. Native species, notably the old English sessile and pedunculate oaks, figure prominently in planting. Equal stress is laid on traditional woodcraft skills: pruning, hand weeding of young trees, and 'brashing', or removing lower dead branches. Operations are labour-intensive, avoiding chemical sprays. And public access and safety must not be an excuse for overtidiness. 'As well as harvesting and marketing woodland products,' the plan notes, 'we are marketing a recreational facility.'

Bristol's urban forest, in short, is a 635-acre enclave in which fragmented identities and land-uses are gathered together and reconciled. Behind the division between the ornamental and the productive lies that of city and country, and behind that, again, lies the green belt, the dormitory town, the second home in the country. Behind the division between amenity and economics lies something more elusive: an assertion that work and pleasure do not mix and must therefore be kept separate. These rifts were made in the cities and it is thus doubly significant that they should be mended there, since for an identity to be successfully reassembled it must be permanently reassembled and this can only happen when it becomes part of the rhythm of daily life. Everything else – the second home, the commuters' village – is a patch-up job, a holding operation.

In retrospect, a form of collective blindness seems to have possessed urban culture during the middle decades of this century. It may well pass for the high summer of industrial society, the first period in history when affluence became not only a mass pursuit but one detached, on a massive scale, from its lines of supply. Ecological considerations – the environment's carrying capacity, the health of croplands and natural resources – were thrust aside in the urge to acquire and consume. In Britain, it can be dated from the end of food rationing to the threat of oil rationing, some twenty years. It assumed its harshest profile in the cities, and it can be partly explained by the fact that by the mid-twentieth century, in sharp contrast to the nineteenth, city-dwellers were no longer hybridized peasants but a pure strain of urbanite.

As the old peasants died out, so did their traditions. Between 1950 and 1970, after the triumphs of the wartime Dig for Victory campaign, allotments halved in number, from 1.1 million to 550,000. The reason, in part, was simple lack of interest, the result of competing new leisure pursuits like cars and television as much as the influence of convenience foods. By the later 1960s, surveys were showing an ageing army of allotment holders. More than four-fifths of them, twice the proportion in the populations, were over 40.

In the 1980s city farmers and urban greeners confirm this picture. The memories, as opposed to the myths, are urban. Techniques of food production have to be learnt anew, traditions created rather than inherited. Hence the cast list changes. The recycler takes over from the rag and bone man, the urban forester from the parks superintendent. The allotment holder, operating a 400-year-old system of holding originally introduced to compensate peasants for the loss of common land, is replaced by the city farmer, who is reclaiming new common land. Superficially their functions are similar; closer inspection reveals them to be, at times, antithetical – just as solar drying is 'merely' a washing line only to those for whom technology is about equipment, not people, and ecology is something to do with wildlife.

The impression is thus of some climactic phase in the history of urbanization and it is surely no accident that versions of the 'ultimate city' – megalopolis and ecumenopolis – arose in response. But if the city had become an accomplished and unavoidable fact, clearly one chapter was drawing to a close and another might begin. The conclusion is hard to avoid that in those mid-century decades a people first savoured fully the experience of a mature urban society and discovered, many of them, that it was not to their taste.

(a) & (b) [top] Greenways and blueways, actual and potential. The route of Leicester's Great Central Railway (a) is being turned into a green corridor through the city. Along the banks of Tyne, a linear riverside park (b) is being created (c) [right] A Town Trees Trust mini-woodland and tree nursery linked to Highshore special school, Peckham, south London

(d) [top] Business park and offices merge into woodland: Lindford Wood, Milton Keynes

(e) [above] The urban village: a neighbourhood park on reclaimed wasteland and demolished housing sites at Elcho Gardens in Calton, east Glasgow.

(f) [left] Footpaths set into a rural landscape matrix, once the site of an ordnance factory, at Warrington Runcorn new town.

LARGER WORLDS:
THE CITY RESHAPED

The millions left after the billions had died tore up the gleaming metal base of the plant and exposed soil that had not felt the touch of the sun in a thousand years. Surrounded by the mechanical perfections of human efforts, encircled by the industrial marvels of mankind, freed of the tyranny of environment – they returned to the land. In the huge traffic clearings, wheat and corn grew. In the shadow of the towers, sheep grazed.

(The fate of the imperial planet-city of Trantor, described in Isaac Asimov's 'Foundation Trilogy')

Manchester was once beautiful. We have this on the authority of contemporary observers from John Leland in early Tudor times to Defoe two centuries later. 'The fairest . . . town in Lancashire,' Leland called it, passing through in 1533. In 1641 the antiquary Peter Heylyn noted 'the beautiful show it bears', while Defoe, in 1727, was much preoccupied with its rivers and streams. The Irwell, he wrote, 'though not great, yet coming from the mountainous part of the country, swells sometimes so suddenly, that in one night's time they told me the waters would frequently rise four or five yards and the next day fall as hastily as they rose.'

It was Manchester's distinctive setting, above all those many small streams feeding down from the Pennine foothills to the north and east into a radial web of river valleys, that led to its foundation. Between the hills and the Cheshire plain, a few miles from the confluence of the Irwell and the Mersey, it was a classic site for a city. The Romans founded a *castrum* at Mancunium on a defensible sliver of land between the Irwell and the Medlock; its remains came to be known as the 'castle in the field', or Castlefield. Defoe put early eighteenth-century Manchester's population at 50,000; 10,000 was a likelier figure.

What happened thereafter is well known. Some time after 1785 the first cotton mill driven by a Boulton-Watt steam engine arrived in Manchester. By 1800, according to John Lord, there were thirty-eight mills in the district. The population was then 75,000. Fifty years later there were 338,000 Mancunians, three-quarters of them born elsewhere, and the beautiful open village had become a place where, in the words of the much travelled and prescient de Tocqueville, 'civilization works its miracles and civilized man is turned back almost into a savage'.

Yet in savagery there was at least vigour. The city's heart beat vibrantly, albeit with a brassy sound. Almost a century after de Tocqueville came George Orwell, to find the vigour dimmed, the people not so much savages as spectres, the place not merely ugly but reflecting a curious desolation of spirit in which slag heaps, chimneys, the 'flashes' of stagnant water in pit subsidence hollows seemed 'a more normal, probable landscape than grass and trees . . .. It seemed a world from which vegetation had been banished; nothing existed except smoke, shale, ice, mud, ashes and foul water.'

Orwell was writing of the 'dreadful environs' of Wigan, which forty years later had proportionately more derelict land than any other local authority district in the country, 30.7 per cent in 1974. It lay within Greater Manchester, which was the English county worst affected by dereliction. The conurbation's rivers were the most polluted in the country, their valleys a collective dumping ground, a vast lumber-room where society stuffed its discards and rejects – anything from domestic rubbish to pylons, power stations and sewage works – and hastily closed the door. Take the River Croal at Farnworth: 'Vast heaps of domestic and industrial rubbish, in places 25 feet high; old coalpits, scrapyards and abandoned factories; glaring white and yellow mountains of toxic chrome waste washing straight into the poisoned river . . ..' Defoe's tempestuous Irwell, meanwhile, had become a foul-smelling old lag, locked away for generations behind grimy factories and warehouses, like an immense and unwanted drain.

But if the life was ebbing from the extremities – by 1982 the North-West as a whole contained more dereliction than any other economic region – the heart was now affected. The city centre, already hit by population and employment outflows, suffered a retailing collapse in the 1970s. Manchester, said a senior council official, was the victim of 'an image problem . . . [It] was seen as a dying city.'

Image problems concern perceptions rather than realities. Public bodies, which deal primarily in the politically saleable, are thus peculiarly ill-equipped to solve them. The boldest initiatives thus came from private individuals or groups whose actions and enthusiasms needed to be justified only to themselves. If Landlife had failed in Liverpool, it would merely have been one small voluntary body the less. For this reason the response of Manchester's local authorities to the image of impending urban death reflects great credit on them.

Plate 11.1 Moses Gate, near Farnworth, Manchester, 1972: disused
mill lodges and rubbish tips

The basis of what has been called the 'Greater Manchester
adventure' was the renewal of charred and rusted landscapes on a
massive scale, even dwarfing for extent and urgency the reclamation
of the Lower Swansea Valley. Between 1974 and 1982, 4,267 acres, an
area over six times the size of the City of London, were reclaimed.
Reclamation, in other words, involved more than three times as
much land as the Swansea Valley and was accomplished in less than
half the time. The nine million trees planted represented almost half
the total for all the provincial metropolitan counties. Ten country and
water parks were developed, involving the most ambitious earthworks
– 50,000 tons of soil to cover the tips and chrome waste around the
River Croal at Farnworth, for example. Here, on land sown with wild
flowers and on old paper-mill 'lodges', or reservoirs, people now
boat, fish, picnic and bird watch in the 1,235 acre Moses Gate
Country Park, a few minutes' walk from the town centre. Such
transformations have to be lived through to be believed.
  The new landscapes were designed to meet particular needs. They

Plate 11.2   Moses Gate, near Farnworth, Manchester, 1982: country
park

were close to people's homes and work; they were informal, even
'rural', and they were planned to attract wildlife. In April 1976
Greater Manchester's wildlife working group, a mixture of conservation bodies and community groups, met for the first time to advise
local authority planners on ecological issues. In 1980 the first
computerized ecological survey of Manchester's wildlife sites was
initiated. Some 270 sites of biological importance were identified
and recorded, to be fed into the land reclamation and planning
process. Two-thirds of these were industrial in origin. West of Leigh,
at Pennington, one of Orwell's stagnant flashes became the focus of a
country park.

In Manchester's other great project, the creation of 'natural'
countryside was paramount. This was nothing less than the stripping
away of more than two centuries of filth and clutter from the river
valleys that had once made the city productive and beautiful. The
principle was the creation of green fingers and wedges radiating out
from the heart of the city along the valleys, widening into country

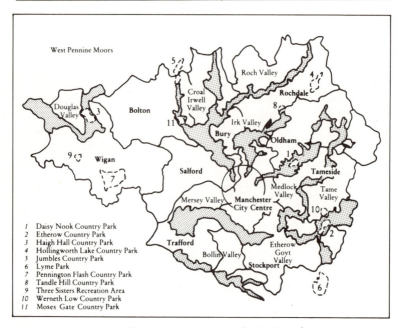

Figure 11.1   River valleys and country parks in Manchester

parks and finally merging with the green belt beyond. Altogether eleven river valleys were involved, under the banner of 'bringing the countryside to your doorstep'.

In the centre of Manchester – 'right at the heart of things,' said the slogan, hopefully – the city was meanwhile sifting through the detritus of centuries, selecting its memorabilia. Around the corner from Ancoats, a curious new landscape arose, buffed and polished, possessing an indefinable air of pastiche. Its components were bundled into a framework of viaducts and canals near the confluence of the Irwell and the Medlock. To the north is the brick-built set of Coronation Street. Further south, in what was once the world's first railway station at Liverpool Road, is the Greater Manchester Museum of Science and Industry. Nearby a Victorian cast-iron market hall houses the Manchester Air and Space Museum and in a handsome old library fronting on to Deansgate is the urban studies centre. Finally a landscape garden was laid out around the north gate of Agricola's fort of AD 79 – the 'castle in the field'. In 1982 the city set the seal on this episode of collective regression by designating the whole area of Castlefield Britain's first urban heritage park.

Manchester in the 1970s was the remains of a gigantic accident, an economic and technological spasm which threw vast numbers of people together, outstripped and exhausted traditional notions of urban design and presented government, such as it was, with one simple but overwhelming task: that of coping. Design, particularly design which encompassed a whole city, was a baroque luxury that could rarely be afforded until growth had ceased. Yet to a greater or lesser extent, Manchester's story is that of almost all other old large towns and cities in Britain. And by the mid-1980s most places had been touched by the environmental renaissance. In October 1983 the Association of Metropolitan Authorities, the cities' voice at government level, was moved to set up its Green Group to examine the extent of the phenomenon. It concluded, in 1985, 'Green policy is widely based ... individual local policies are developing despite the absence of a coherent national policy.'

The AMA's enquiry also inadvertently revealed a general haziness about the nature of green policy. Anything from a nature reserve to a ski slope or a mining museum could, it seemed, be regarded as

Plate 11.3 The 'castle in the field': the replica Roman fort now part of Manchester's urban heritage park

'greening the city'. Struggling to extract some coherence from diversity, the AMA provisionally set down the wider aims of local authority green policy as follows: enrichment of life; mental health and well-being, and increased economic activity arising from better working conditions. The theme common to them all seemed to be the pursuit of pleasure in a park-like setting, a vision Arcadian rather than urbane.

Of central importance was a massive change in land-use which has passed largely unnoticed. In the West Midlands, for example, of 1,820 acres of derelict land reclaimed by the county council in the decade from 1974, forty-five acres were destined for housing, 115 acres for industry and 1,660 acres – over 90 per cent – for open space. Of 1,726 acres restored by Tyne and Wear county council between 1974 and 1983, 1,490 acres, or 86 per cent, were for recreation (56 per cent) or 'amenity/agricultural' (30 per cent). In a typical Inner London borough like Newham, open space doubled over the same decade. In Sunderland recreation and open space by 1985 represented 'one of the most extensive urban land uses'.

Closer analysis, moreover, shows that the bulk of all urban reclamation, more than four-fifths in the case of Tyne and Wear, involves a conversion of land from industrial to recreation, amenity or agricultural use, a post-industrial transformation in its most literal sense. Applying the pattern for Tyne and Wear to all seven English conurbations between 1974 and 1982 would involve some 13,000 acres of industry, equivalent to 7,300 Wembley football pitches, being knocked flat and replaced by grass and trees. This was urban land-use change on a scale probably unprecedented since city boundaries stabilized.

The scale of reclamation was matched by increasing urgency – 30 per cent faster in the second half of the 1974–82 period – and developing expertise. The climax was the 1984 Liverpool garden festival, the fashioning, in thirty-one months and two growing seasons, of stream, lake and forest, an 'allegory of a mountain range', out of 250 acres of silted-up dock, oil terminal and rubbish tip. Altogether over three million tonnes of rubbish and clutter were moved, a quarter of a million trees and shrubs planted, a hill raised 140 feet above the Mersey. The hill is made of domestic refuse, capped with clay. From its summit one can see, on a fine day, the mountains of Snowdonia.

The international garden festival was the biggest single inner-city reclamation project ever attempted in Europe. It was probably the most concentrated large-scale piece of reclamation and landscape design in history. But beyond these superlatives were more significant and lasting achievements. At Liverpool men took a working landscape and turned it into one of pleasure. They found a ruined slice of a spent city and made a garden. They proved that asphalt was not necessarily 'the land's last crop'. These things were known and done before the garden festival, by all the small-scale makers of pocket parks and nature gardens. At Liverpool they lodged firmly in the national consciousness. Garden festivals have become national policy – Stoke in 1986, Glasgow in 1988, Gateshead in 1990 and a Welsh city in 1992. Urban reclamation has gained enormously in pretensions. The £1 billion 'water city' in London's Royal Docks, taking in ten miles of quayside and 700 acres of land, is destined to be Britain's most ambitious single piece of urban renewal.

The Liverpool garden festival may prove a watershed in several important respects. It was a collective recognition, somewhat belated, that urban regeneration had assumed a deepening greenish hue. It was a long-overdue reinstatement of the landscape architect in a position of professional primacy. It marked a diffusing awareness of land as a renewable resource, and hence of the value of earthworks to a society in which rapid technological change was bringing rapid land-use change. Above all, perhaps, it was an acknowledgment by central government that derelict landscapes make derelict people – and hence that the opposite may also be true.

As the following pages show, these themes surface repeatedly in a host of disparate spheres and developments, from the design of offices and the location of industry to the renewal of London's docklands. The garden setting, the green framework, becomes pervasive, insistent. 'Park', as in the ubiquitous science park, becomes a form of invocation, a leitmotif of promotional imagery. There is a sense of a culture reaching out to something, some ideal or lifestyle, which it has not yet properly defined. 'Environmental quality' is a poor term to describe it.

A starker one is the 'economics of amenity', the name of the programme launched in the United States in 1980 by the nationwide voluntary organization Partners for Livable Places, and which has been adopted in more than thirty cities. The programme defines three types of city asset, cultural, recreational and natural, and practises 'quality of life planning' as a means to economic renewal. It is based on sober analysis of the factors influencing the location of homes and businesses, and builds on some impressive case studies in which cities have pulled back from the brink of dissolution. They have done this, not to put too fine a point on it, by making themselves green, beautiful and entertaining. Such initiatives are best summed up as an exercise in collective gentrification – 'enveloping' on a grand scale in what is usually styled a landscaped setting. Out of scores of examples, two stand out: the old New England city of Boston and nearby Lowell, a kind of transatlantic Manchester. In Boston, Faneuil Hall marketplace, consisting of three early nineteenth-century buildings, was restored and developed, turning into a lively and thronging centre of small shops, market stalls and eating places. In Lowell, the empty mills and warehouses with their network of canals became in 1974 the first urban heritage park and, in 1978, a national historic park.

For the governments of expiring cities, such dalliance with environmental determinism was a journey to the end of the rainbow where lay, possibly, a crock of gold. The lure was tourism and its revenues. By 1981 twelve million people were coming to Boston's Faneuil Hall each year. Half of them were visitors. In Lowell,

Plate 11.4   The city as playground: Bristol's revitalized docklands

unemployment dropped from double to half the national average in four years. Other American cities thus began feverishly to upgrade their 'livability', launching renaissance projects, creating parks, initiating greenway systems. The Boston model was meanwhile transferred to London's Covent Garden, a fitting place, since it was here, three and a half centuries earlier, that the Earl of Bedford and Inigo Jones gave Britain its first taste of modern city planning, in the shape of a high-class residential area based on the Italian piazza. Thus was born the square, later acquiring trees, which was to contribute so much to the surviving habitability of the city during its industrial eclipse.

Thanks to some assiduous prompting by the English Tourist Board, the early 1980s saw the message spread to British cities. In London, Camden Lock and St Katharine's Dock followed Covent Garden. Bristol's docks spawned an arts centre – the Arnolfini Gallery – an exhibitions hall, a camping and caravanning site, a water leisure centre and a maritime heritage centre. Linking them was a waterside walk itself forming part of a long-distance footpath, the Avon Walkway. The docks burgeoned with new activities: regattas, power boat races, firework displays, the World Wine Fair. In the quest for urban distinctiveness, meanwhile, the most sulphurous of histories was little obstacle, since with it came, usually, a wealth of age-encrusted artefacts and saleable associations.

The result was a rich seam of paradox as some of Britain's most workaday places, many of them bywords for industrial gloom and despondency, attired themselves in new and verdant livery. Bradford, sprouting wine bars, a 'Green Plan' and a basketball team called the Mythbreakers, sold its cloth caps, clogs and newly pedestrianized cobbles in the form of packaged holidays: tourists came to view 'Mill Shops' or the 'Industrial Heritage'. More significantly, it promoted 'psychic sightseeing' – tours of Celtic heads, runes and Dobby stones, the latter prominent in elven lore. Wigan, unabashed, staged a 1984 George Orwell festival, complete with a specially brewed Big Brother ale, to celebrate the renovation of its pier. Holidaymakers visited Scunthorpe, water-skiers and wind surfers colonized London's docklands. Halifax, which opened its Piece Hall, a neglected prototype for Covent Garden, in 1976, was by 1985 toying with *A Strategy for Prosperity*, a report recommending a comprehensive facelift centred on a major new urban park and a cable-car system covering much of the town centre.

The artefacts meanwhile grew even larger. If Battersea power station was to become a fantasy park, its main hall themed to resemble pre-industrial London, Manchester, not to be outdone, proposed turning its ship canal, the deepest in the world, into a centre for sailing, water-skiing, power-boating, sailboarding and hovercraft. By imperceptible degrees the city was being turned into a gigantic playground and an open-air museum.

In a related process, the most resolutely functional of developments acquired a green patina, a kind of afterglow symbolizing, evidently, an attempt to harmonize work with pleasure. Pleasure in this sense, however, was a shorthand for values regarded, hazily but genuinely, as both indispensable and life-enhancing. Hence the science park, a post-industrial setting for post-industrial technology, and its cousins: the business park, the venture park, the high-technology park. After a slow start, Britain caught the science park fever, from California by way of Cambridge, in the early 1980s. By 1985 twenty-two science parks were in being or projected. The spur was the Cambridge Phenomenon, the economic growth associated with East Anglia's silicon fen, but there was a deeper message. High-technology folk, said Mr Henry Bennett, manager of the Cambridge Science Park, 'appreciate pleasant, spacious and restful surroundings and wouldn't go to an industrial-type estate at any price.' In 1985 a study by Cambridge University geographers proved him right. The most striking clusters of high-technology firms, it found, 'have evolved in residentially attractive areas possessing existing major research facilities.'

Where high technology led, offices, services and retailing followed. At the forty-acre Kembrey Park business centre at Swindon, courtyards, pergolas, grassy banks and tree-lined walkways are complemented by attempts to create a sense of place through landscape structure and planting. At Linford Wood business centre, in Milton Keynes, native planting, centred on a pergola and tree grove, aims at an ecological relationship with surrounding woodland. Commercial property generally has seen a revolt against the 'let and forget' syndrome, and the rent-collecting slab it produced, and a new emphasis on quality of design. In office buildings the atrium, or inner landscaped garden, has proliferated. 'Campus-style' developments are increasingly popular.

In all these developments, 'image' is clearly a crucial factor. Indeed, in terms of the outflow of jobs and people from cities, it seems inescapable. We are, in a sense, pursuing dreams and fleeing nightmares. And despite almost three decades of research, the fundamental statement on image remains that of Kevin Lynch in 1960. This was a city consisting of five types of form, the path, the edge, the district, the node and the landmark, all of which, depending on their configuration, made a city more or less 'imageable'. Lynch's 'image' is closely related to the mental map of Peter Gould, the 'schema' employed by Terence Lee, the 'cognitive domain' of Amos Rapoport. It is a way, indeed, it is *the* way, by which the individual shapes and classifies experience to give it order and meaning. As such, it is clearly critical to his identity: way-finding is the image's original function and Lynch cites several memorable examples of the ontological terror engendered by disorientation. A city with a vivid and articulated image, imprinting itself clearly on the mind, is not only efficient and accessible. It also has form and character, both aspects of wholeness, and can thus become an object of beauty or affection.

Lynch's achievement in *The Image of the City* was the invention of a vocabulary of urban design based on the responses of those who used cities, rather than those who traditionally planned and ran them. *Imageability*, it seemed, was a poor substitute for *visibility*, but one made necessary by the sheer scale of the modern city. Florence, set amidst hills, centred on the Duomo, bisected by the Arno River and penetrated almost to its heart from the south by open country, was Lynch's model of the visible city. It could be viewed 'whole'. But cities on a metropolitan or regional scale were imageable only with extreme difficulty. A dominant landmark, a river or a mountain, might help, but topography – what Lynch called the 'pre-existing natural setting' – seemed obsolete as a determinant of image, overwhelmed by urban size, density and technology.

Many of the ideas expressed by Lynch and succeeding imagers of the city were clearly at work, as the opportunities for redesigning cities opened up over the following decade. 'The city should ... speak for itself,' declared the plan for brave new Milton Keynes, in 1970. 'People should be able to acquire a clear working knowledge of the city and its form through direct experience.' But creating an intelligible form on the empty canvas of green field Buckinghamshire was much less complex than distilling order out of ancient jumble.

This was the challenge, the widely shared perception of the city-as-industrial-accident and the urge to clear it up, which helps to explain the growth of the heritage and urban studies movements. By 1985 there were forty heritage, formerly architectural interpretation, centres in Britain, two-thirds of them in towns or cities. They served in all a population of about two million. Their roots lay in the dramatic rise in interest in local history dating from the later 1960s – the number of local history societies, for example, doubled between 1970 and 1976.

The first urban studies centre, meanwhile, opened in 1971. By 1984 there were forty, mainly in the larger cities. Both heritage and urban studies centres owed their existence largely to volunteers; both sought, through explaining and interpreting the city, to *connect* people to it, and thus to give them a sense of comfort, identity and belonging.

Birmingham, experimenting with community participation in planning, meanwhile witnessed a novel attempt to confer the cognitive freedom of the city on its inhabitants. One of the surest proofs of Lynch's theories is the behaviour of virtually every visitor to a strange city. He searches out the centre to ascertain its vitality. He also climbs the high points to judge its form. In general, however, urban vantage points tend to be accidental or expensively private. At Birmingham, on a man-made hill at the centre of one of the city's most ambitious reclamation projects, the planners have designed a toposcope, a 360-degree plan which, when held at eye level and correctly oriented, fuses map, reality and perception in a unified image of the city and its landmarks, from the old Singer car works a few hundred yards away to Frankley Beeches nine miles distant.

The project is known as the Ackers, forty-five acres of reclaimed dereliction less than two miles from Moseley Bog. The Ackers lies near the confluence of the River Cole and the Spark Brook and is crossed by the Grand Union Canal. It is thus a vital staging-post both in the West Midlands' strategic wildlife network, here represented by the River Cole corridor, and in its canals' network. In terms of contemporary ideas on urban design, it is a much-favoured position.

Greenways and to a lesser degree blueways – linear space systems based on rivers, canals, parks, disused railways, footpaths, bridleways and any other serviceable and coherent open landscape features – are fast becoming the leitmotif of large-scale urban restructuring. They are seen as routes for a special kind of traffic: traffic which passes peacefully through the city employing technologies appropriate to that purpose – cycling, riding, walking. They are ways of pleasure rather than of work – byways, not highways. And they are visualized as thronging with wildlife, which it is their purpose to lure into the city and disperse throughout it.

Major cities practising a recognizable greenway strategy include, as

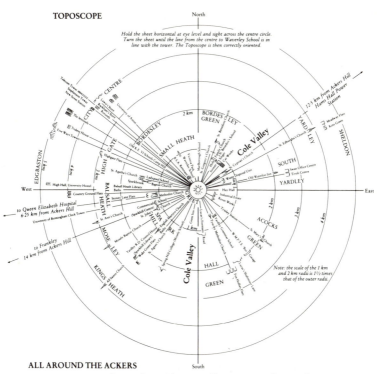

Figure 11.2   Views from the Ackers Hills, Birmingham: the toposcope

well as Manchester and Birmingham, London, Glasgow, Liverpool, Cardiff, Swansea, Newcastle, Sunderland, Sheffield, Coventry, Leicester and Stoke. Among the most interesting is Leicester where Great Central Way, once a railway line following closely the course of the River Soar, will bisect the city from north to south when completed, providing a traffic-free route to the centre. In Newcastle the Network project is planned to cover 200 miles of greenway linking city with sea and moorland, connecting parks, picnic sites and wildlife reservoirs. An even larger-scale example is the southern half of the North-West economic region. Here, in the black belts between Liverpool and Manchester, the Groundwork scheme is creating a lattice work of greenways around towns like St Helens, Wigan and Macclesfield, linking up with their big city neighbours to east and west. Groundwork, a partnership between local authorities, private companies and volunteers, was begun in 1981 and launched nationally in 1985.

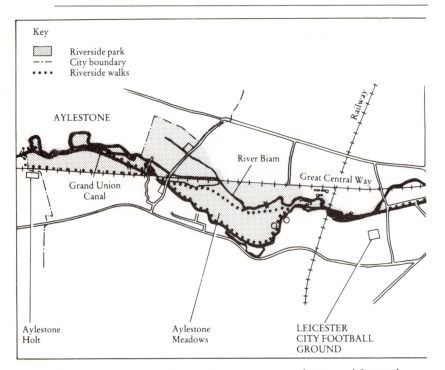

Figure 11.3   Riverside corridor, Great Central Way and footpath system, Leicester

The greenway strategy is in one sense a re-embodiment of Ebenezer Howard's social city. The greenways serve to 'village' the city, fragmenting an unmanageable abstraction into small habitable neighbourhoods, separated and identified by swathes of green. They are thus part of a tradition which takes in half a century of neighbourhood unit planning as well as the original objectives of green belt policy.

Two other persistent themes of contemporary urban design, the preoccupation with the image of the city and the quest for an ever-larger parkland setting for housing and employment, owe much to this tradition. In the Greenline environmental improvement policies for road and rail corridors throughout the West Midlands, for instance, 'image' has clearly assumed regional dimensions. Both themes effectively involve 'enveloping' on a grand scale, an attempt to dissolve the city by removing its encircling boundaries. City and country interfuse and merge, as Howard envisaged.

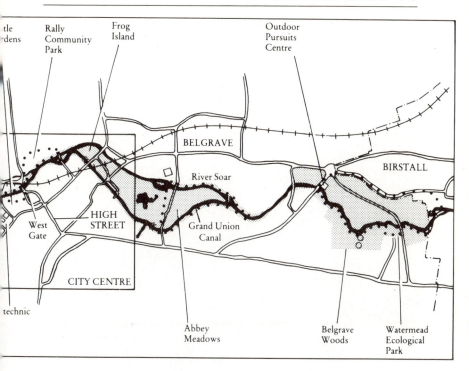

The older traditions of urban reformism thus shed much light on the greening of the cities. Their most obvious lacuna, however, is the rise of ecology. And after a hesitant start, large-scale land agencies, notably local authorities, have responded with unaccustomed alacrity: the late 1970s appears to have been a turning point.

The timing of this conversion has much to do with unemployment, national habitat loss and the rhythms of urban policy. But there are more specific factors. British local government reorganization in 1974, for instance, gave many councils their first ecologist. Most were young and straight from university and their role, initially, was defensive. Circumstances soon changed, however. The first serious local authority spending cuts began in 1976. These coincided with the run-down of many of the elaborate Victorian parks – Battersea and Victoria, in London, are well-known examples – which were proving prohibitively expensive to maintain. The prospect of vast new tracts of recreation space made a formal high-maintenance

Plate 11.5   Villaging the city: private gardens and leisure gardening plots have given a village atmosphere to an area of streets and demolished housing at Elcho Gardens in the east end of Glasgow at Calton

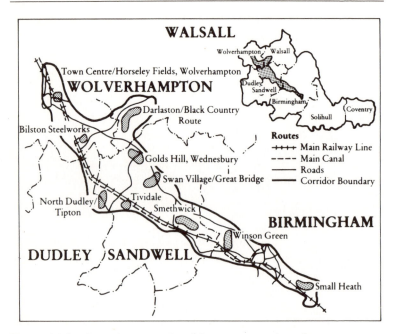

Figure 11.4 Greening as regional image promotion: the Birmingham to Wolverhampton corridor initiative, part of the West Midlands Greenline project, involves environmental improvements along key transport routes

landscape even more unrealistic. Accumulating evidence of widespread public preference for informal recreation also seemed to point towards a 'natural' setting.

The new landscapes advocated by the small-scale city greeners had the inestimable benefit of cheapness. Volunteers or Manpower Services Commission workers made them, ususally with ready-to-hand materials. Maintenance costs were low. Tulse Hill nature garden, built by children in Brixton, cost £2,000; an equivalent Greater London Council scheme would have cost £170,000. Native semi-natural planting can cost less than a tenth of orthodox grass with trees and, somewhat surprisingly, a fifth, in both laying and maintenance costs, of concrete on hardcore. At Warrington new town, ecological planting costs, even without voluntary labour, are put at between a tenth and a quarter of conventional methods.

It is instructive to ask why the new landscapes are cheaper. The fundamental reason is that they mimic – they are *designed* to mimic – the organization of natural communities of plant and wildlife. Over many thousands of millennia these ecosystems have evolved habits of thrift, of exchange and recycling of solar energy and nutrients, which make them self-sustaining.

Yet this masks a deeper truth. It is a relatively easy matter, given abundant supplies of energy, money and machinery, to chop, mow, spray, plough and poison. It requires infinitely more patience and skill to tease out and re-create the endless intricacies of a self-sustaining ecological community. In 1972 permission to extract gravel from Staines Moor in Surrey, Britain's first notified site of special scientific interest, was refused. Ecological knowledge was simply inadequate to restore the site's species-rich grasslands after use.

Since then expertise has grown enormously, partly as a result of the challenge posed by dereliction. Ecologists in Manchester have grown plants on raw, highly acidic colliery waste, without using topsoil. Terry Wells, who has pioneered 'artificial' meadow creation at the Institute of Terrestrial Ecology, believes it is now possible to make good, if not exact, copies of ancient grassland. Wells's view is apparently shared by the growing number of public bodies which are sowing parks, road verges and factory sites with wildflower mixtures. At Liverpool University Tony Bradshaw and his team have developed techniques for growing grass on bricks. Society may soon be able to create habitats as successfully as it can now destroy them.

The ecological spirit has produced small-scale convulsions in public practice that would have once seemed heretical. In Birmingham the casual term 'urban desert' now denotes an officially sanctioned wildlife action area, existing 'where accessible habitats with wildlife interest are located more than a kilometre distant'. The county altered its structure plan accordingly, espoused a recognizable version of bioregionalism, set up a wildlife records centre, endorsed wildlife gardening and declared in its wildlife policy document, 'We need nature conservation for our own good.' Huddersfield has taught its parks maintenance staff an entirely new language, dealing in irregularity, randomness, drifts, glades, mosaics and scalloped edges. A council-run ecological advisory service was set up in South Yorkshire, post-Orwellian Wigan launched, in 1984, its own nature conservation strategy, and London's first local authority ecology department was established by the Greater London Council. The Greater London development plan was also changed. 'A rich and varied wildlife community of plants and animals,' the new version declared, 'is an essential part of Londoners' environment.' By the mid-1980s most major cities had conducted an ecological survey of their trapped and unofficial countryside, often storing the results on computer and feeding them into the planning and development control process.

These examples give only a flavour of the injection of ecology into the formal planning of cities. Yet set together with greenways, with the rediscovery of urban topography, particularly rivers, with the spread of environmental determinism and the emphasis on the parkland setting, they have all the ingredients of a new design philosophy. It is as though another type of city, an alternative or post-industrial city, is striving for expression, its shape visible through the outlines of the old like an insect larva in the final stages of pupation. This rival city has its own transport network, its own industries, its own ways of taking pleasure. At present the two cities, the industrial and the post-industrial, co-exist but the balance is unstable. The industrial city clearly has much more shrinking and unravelling to do.

A conspicuous failure of the modern city is that its buildings lack relation to one another and to the landscape. The release of space and the recovery of natural form, of hidden rivers and man-made hills and woods, offers the chance to restore lost connections. In his study of Greek sacred architecture Vincent Scully showed how an intense devotional feeling towards a particular grove or mountain – a feature which did not 'represent' a god but embodied him, made him determinate – resulted in an almost intuitively harmonic relationship between landscape and built form, often structuring a large area. An example is the sacred processional route associated with the Eleusinian mysteries in Athens. Another is the palace of Knossos in Crete which 'became [the] body' of the great mother goddess, 'as the earth itself had been in the Stone Age'. A clue to this achievement was the Greeks' occupation of a moment in history when the deepest beliefs of the Palaeolithic ages came into harmony with a 'new and liberated thought'.

The role of the Panathenaic processional way in creating connections and relations between buildings and thus in shaping a city is also cited by Edmund Bacon in *The Design of Cities*. It was not only a 'central spine' of Athens's commercial and political activities but 'part of a system of regional movement which linked some of the most sacred places in Greece'. Such a clearly expressed movement system is a powerful influence 'capable of seizing men's minds and developing loyalties around it. Of itself it becomes a major political force.' Bacon goes on to speak of the basic design structure as 'the binding together of perception sequences shared by large numbers of people'. This is to be achieved, he suggests, by strong articulated nuclei based on beloved landmarks distributed throughout the city and related firmly to natural features and regional topography. These two authors writing in the 1960s – one a Yale professor, the other a leading American city planner – clarify the fundamental relationship between the form of cities and the structures of religious and political belief. They thus offer important clues to the shape and function of the emerging alternative city. One thinks of Birmingham's toposcope or of the environmental trails proposed in an ecological study of Glasgow where spines and shoulders of land would provide 'important vistas over wide areas of the conurbation . . . imparting considerable geographical perspective'. One thinks of the West Midlands' conurbation cruiseway and its strategic wildlife network, in which the movement is along routes shared with wildlife and articulated by places devoted to it: articulated, too, by places of history and recreation.

One thinks also of the 'green chain' of open spaces in south London or of the green 'necklace' initiated in 1979 by the Greater London Council along the Grand Union Canal and composed in part of painfully constructed neighbourhood green patches like Meanwhile Gardens. Was this union of grand urban design and small local initiative – co-operation, in Jamie McCullough's words, between the elephant and the flea – a happy accident or did it indicate some deeper shared vision of the city? Is it significant that, like Alfred Watkins's ley-lines, there is a curiously mythic quality about the greenways – that they relate less to an observable reality, which is frequently anything but green, than to a sense of the richness and mystery that may lie behind it? We label them, antiseptically, 'urban planning concepts'. Are they not closer in spirit to the legendary 'dreamtime roads' of the Aborigines of central Australia? And if green spines and shoulders are the bone structure of the emerging city, might not rivers – an 'important character-making asset', according to Cardiff planners – be its soul? Might Manchester's imprisoned Irwell, if not returning to its original tempestuousness, at least regain an independent spirit? Ebenezer Howard's vision, in that case, would need modifying – a garden city perhaps, but the garden is wild.

These are merely hints and suggestions, but if the physical creation of the earth-shaped city owes much to applied ecology, particularly that involved in landscape design, its inspiration remains the 'secular religion' of environmentalism, the most widely diffused mode of spirituality in the post-Christian west. And in the recovery of the primitive associated with environmentalism one can recognize another moment in history, like that in classical Greece noted by Scully, when 'liberated thought' is seeking a balance with atavism and folk culture. Unlike priests and kings, however, environmentalism has little experience of making cities: this was not why it came into being. It has simply been forced to confront them. In that sense the greening of the cities is a great adventure – a journey, very much, into the unknown.

12

CONNECTIONS AND RECONSTRUCTIONS

Having lived among such deprivation for the last 50 years, the local people have lost confidence in themselves . . .. Anyone with get-up-and-go has got up and gone . . .. This lack of leadership quality makes obtaining the services of men and women able to lead any kind of organization a long, wearisome and often hopeless task . . .. Constant harassment has forced a very kind and able parish priest to quit his inner-city church . . .. I am sure that he, like myself and many others, feels like a nurse looking after a terminally ill cancer patient.

(Vicar of Anglican parish in inner Salford, quoted in Guardian, *14 February 1984)*

The Factory has now become an impressive community resource with a coffee bar, youth centre, craft room, toy library, toddlers' play area and adult education classes . . .. The outside walls have been transformed by a colourful mural, portraying the edges of a town turning into countryside.

(Conversion of an old piano factory, near the Ball's Pond Road, east London, into a community centre by voluntary groups, described in Waking up Dormant Land, *CoEnCo, 1981)*

He said, 'Look, you don't want shrubs or plants, you'd be better off with something that's easy to maintain and clean. We've got some really nice coloured tarmac,' he said. 'You can have green or red.'

(Liverpool council official, quoted by members of the Weller Street housing co-operative in Toxteth)

A rival city needs people to build it, many people. Those people need vision, opportunity, a space to experiment in, but above all they need power and money. From 9 October 1975 these were supplied in unprecedented abundance.

On that date the newly established Manpower Services Commission launched the first of its special job creation schemes as a response to the rise in unemployment beyond the million mark. Two years later the 'official' urban programme, begun in 1968, was at last extended to cover money for environmental projects, an emphasis reinforced by new guidelines issued in July 1981. The Conservative Government of 1979 also laid greater stress on the private and voluntary sectors. But the scale of official urban aid and the activity it generated was dwarfed by the funds made available through the job creation programme, and its successors, the lion's share of which has gone into the urban areas. Between 1975 and 1984, for example, £1.35 billion was spent on the urban programme. Comparable expenditure on MSC special employment measures totalled £3.3 billion.

The unintended consequence was a significant surrender of power by government, defined in its broader sense as the spectrum of public authorities and official agencies. The framework evolved for job creation recognized that, in generating and matching up a multiplicity of local employment needs and opportunities, the centralized apparatus of government would be too cumbersome. Thus was born the 'sponsor' body. Sponsors could be local authorities, voluntary organizations, trusts, charities, private firms, *ad hoc* groups, even individuals: the rules stipulated merely that they 'should have the resources and capability to undertake the project'. The project itself was required to provide 'worthwhile work', not viable without job creation funding. It should 'make a contribution to the enhancement of the local environment or assist in the solution of a social or community problem'. Urban renewal was a priority.

The impact proved considerable. Hundreds of community groups learnt new skills of project and people management, a great gain in autonomy and self-confidence. The needs of a community have been defined by the people living in it, who have also been given the chance to act, and to act quickly. Those with hopes, or visions, for a community have been given freer rein to express them.

The diffusion of power and resources to the local and the unofficial entailed in job creation was without precedent. It also proved vital to the greening of the cities. It allowed open-minded local authorities to experiment, to escape the webs woven by their own bureaucracies and trade unions, to pursue an image outside the constraints of their budgets. For the legions of unofficial greeners it often made the difference between extinction and survival.

In the broadest terms a perceived crisis – in this case, one million unemployed people – was to prove the catalyst for fundamental social change, panicking government into suspending its rules and loosening its grip on the levers of state control and expenditure, a step which was consciously repeated later in experiments with enterprise zones and freeports and in the much wider debate on deregulation and privatization. The fact that many of these developments originated in the cities, or evolved out of concern for them, is no accident: this was where the obsolete clutter, political and organizational as well as physical, was concentrated. A culture has to be reworked where it has most signally ceased to work. But the wider lessons should not be obscured. The dismal record of inner-city local authorities in dealing with decay led finally to the setting-up of urban development corporations, taking over many of the councils' powers. Yet the failure of inner-city democracy could not be an isolated phenomenon, somehow detached from the local and national democratic system of which it formed a part. What broader failings were there? And what new approaches were being tried?

There is an added significance in the crucial role played by the 'unofficial' urban policy in the redesign of cities. The surge in unemployment which prompted it marked the transition to a post-industrial society in its most unexceptionable sense, that of a shift from manufacturing to service- and information-based work. It also sharply accelerated the previously even-paced decay of the human values associated with industrial society. Cities were shaped and built to suit, broadly, an industrial work ethic. It was thus ironically appropriate that the response to a devastated industrial employment base should finance the reshaping of the city in the service of a revised work ethic – or, perhaps, of no work ethic at all.

By the early 1980s a new consensus was emerging. Many centralist functions of government appeared largely inoperable. Orthodox state funding of welfare-style programmes gave temporary relief but ultimately increased the patient's dependence. Various forms of prodding, prompting and pump-priming were thus essayed. Regeneration remained the nominal objective of public policy but it struck no popular chord since it was too often interpreted in practice as meaning that cities should continue to be what they had always been whereas most people wanted them to be different. Against the revealed impotence of the central state, meanwhile, were set ideals of neighbourhood development and community self-reliance which were new and still relatively untested. The state, in other words, was looking for answers from the neighbourhood, answers that would almost certainly prove crucial to the future course of democracy.

Central to the main reformist critique of the city is the loss of community. In this tradition, an imagined pre-industrial world characterized by smallness of scale and close family and kinship ties is contrasted with the age of rootless, fragmented anomie produced by industrialization. These two extremes, characterized by Ferdinand Tönnies in 1887 as *gemeinschaft* – roughly, 'community' – and *gesellschaft*, and elaborated by Durkheim and Georg Simmel, lent a permanent structure to sociological thinking. Industrialization, coupled with the factory system, the separation of work from home, and of both from the land, led to lives being lived in separate compartments. Hence, according to Max Weber, China's failure to industrialize: it resisted the division of labour and 'rational' management techniques because of religious and cultural traditions which stressed the well-rounded human being living in harmony with the cosmos. Hence, too, the concept of the specialist city, pigeonholing people and treating landscapes as filing systems, which was must fully articulated in the human ecology of E.W. Burgess and Robert Park. Invoking the rhythms of plant and animal communities, the Chicago school described a struggle for space in cities. City centres were the captives of business but around the central business district lay a series of concentric zones containing different land-uses and human communities: 'hobohemia', slums, the Latin quarter, the 'bright light area'.

A corollary of this argument is that the dismantling of industrialism will lead to a reassembly of identities and that this, in turn, will be reflected in land-use. The city farm is probably the most notable example of re-integrated selves jostling for multiple expression through a single plot of land. But there are many other signs of a slow revolution in land-use. There is the rise of the household economy and 'home-centredness', the growth in self-employment, home-working and the DIY industry. There is the rapid spread of 'employment accommodation', combining industry, offices and services under one roof. New high-tech industry, seeking a rural setting, takes over old barns and farm outbuildings; factories assume a vernacular look. The relationship between work and free time is being fundamentally revalued. And on a larger scale the segregated land use of industrial urbanism faced irresistable subversion.

Is a science park, for example, 'industry' or 'open space' or even, as in the case of Linford Wood at Milton Keynes, 'forestry'? If rural agriculture, overtly industrial in tone, is free from planning controls, should not city farms, overtly anti-industrial in tone, enjoy the same privilege? When, as in the case of Bristol, can one say an amenity woodland has transformed itself into a timber plantation? High-technology industry, kinder to the environment than its predecessors, has served notice that long-established criteria for deciding who occupies which bit of land for what purposes, criteria enshrined in the use classes orders of town and country planning legislation, are increasingly obsolete. But this merely mirrors a far more profound disruption of living patterns.

Daniel Bell has argued lucidly that cultural forces have now replaced technology as the dominant agent of social change. But one does not need to endorse Bell's thesis fully to see that these land-use transformations are ripples from a cultural wave. Behind them lies the post-industrial quest for self-expression and personal wholeness. One can dignify it as holistic or stigmatize it as hedonistic, narcissistic, endlessly concerned with self, with self-analysis, self-projection, self-fulfilment – Tom Wolfe's 'me' generation. Its concern is to be multi-faceted, to avoid categorization – not man or woman but 'person'. It wants to be rich *and* beautiful *and* fit. Its emphasis on individualism – all that remains, according to Bell, of the Puritan ethic when finally stripped of any pretence at the transcendental – gives it a rebellious, anti-authoritarian spirit, a tone that insists on rights and entitlements. Its doubts about 'standards', its awareness of a multiplicity of styles and opinions, find expression in a certain easy tolerance and informality. By and large it resents and distrusts ideologies, establishments and politicians. Its chief dilemma, perhaps, is to discover at what point self-exploration and self-reconstruction become self-indulgence.

Inevitably, this composite of desires and attitudes seeks a scale and a setting appropriate to its consciousness of enriched identity. It is in the more or less conscious search for the appropriate setting – a quest with many precedents in the history of landscape symbolism – that the urban greening of the 1970s and 1980s plays so important a role. How, in simple terms, must the look and feel of cities alter to express the sea-change in outlook and activity of their inhabitants? Historically, however, it is around the theoretically more amenable question of scale that the debate has crystallized.

Fundamental to the 'community lost' tradition of anti-urbanism is its recoil from the sheer size of the modern city. In practical terms, as we saw, urban growth produced bureaucracy and professionaliz-ation. Power was pushed upwards to larger, more 'logical' units. Localism withered. In human terms the arguments centred nominally on urban-rural contrasts – crime, mental health, disease and mortality – which usually worked to the disadvantage of cities, and on linked questions of crowding and deindividuation. Thus was born the 'non-person', the faceless urban functionary, and the 'co-present other', the anonymous fellow tube passenger or city-centre scurrier. One learned to cope with crowds of such people, it was argued, by a 'decrease in affiliation', by treating others as if they did not matter.

Such responses had the effect of discounting large areas of human experience. Decreasing one's affiliation to others, treating them as non-persons or co-present, was a necessary evil of urban life. Moreover it was 'merely' temporary. Ultimately, somewhere else – at home? on holiday? in heaven? – both we and they could get back to the business of being real, 'whole' persons. In the meantime, however, a certain amount of mutual beastliness was perfectly permissible.

The logic, it will be seen, is that of the industrial work ethic, notably its transcendental component. Present place and time is variously postponed, endured or ignored because other places and times await. And the industrial city, built to service God and Free Trade and combining land-use segregation and green belts with unprecedented size, distributed people over vast areas, miles from their workplace, and opened up large lost and empty spaces – represented, for example, by the daily journey to work – in their lives. Thus arose those armies of the co-present known as commuters who are, in a sense recognized by cognitive psychology, blind to their surroundings. The commuter is, by definition, always *in between* places and times, and thus unconnected to the uniqueness of either. He is, in fact, in a psychological kind of no man's land. From here it is a very small step indeed to the 'unreal city' of T.S. Eliot's waste land and modern urban myth.

Yet the 'community lost' debate was, even more fundamentally, another version of that great intellectual stand-off between the holists and the reductionists. The city's defenders came not so much from the tradition which saw urbanism as liberating or civilizing as from the dominant school of positivism which simply refused to countenance large and comprehensive statements about 'the city'. Broader measures of urban behaviour – concepts like density, anonymity, deviance – were thus broken down into a series of increasingly fine-grained indicators to the point where the city became a lumpen aggregate of a myriad individual interactions from which only the foolhardy and the unscientific would seek to extract much overall pattern.

To the decay of transcendence, however, has been added the retreat of reductionism. Systems and organization theory, meanwhile, have underpinned and refined older notions of organicism. The late twentieth century can thus explain to itself, in terms simply unavailable to previous generations, how a city, part system, part organ, part organization, can develop a momentum and a 'mind' of its own, quite independent of the people who make it up. In this respect it is analogous to the nation state, which it was so instrumental in creating. Both are man-made super-organisms which behave holistically, in ways not readily explicable in terms of the relationship of their parts.

Human fear of the giant industrial city thus becomes a form of species conflict. For man to assert his wholeness and autonomy, he must disrupt the vital functions and patterns of organization by which the super-organic city sustains itself and which condemn him to a fragmented existence. To that extent, his assault is political and social: these are the words we use to describe those vital urban functions. But since the city has an oppressively corporeal existence, he must also destroy its form and fabric, tearing himself out of its body like a renegade sub-organ. In this sense, he is required to be an earthworker and a landscapist. The city is among the most powerful super-organisms men have invented, more so than the nation state since, unlike the latter, it is physically omnipresent, enveloping us in brick, glass and concrete. Dismantling it is thus an imperative of post-industrial culture, with its emphasis on what Alvin Toffler has christened demassification.

New intellectual disciplines have thus enormously clarified the issues at stake. They also offer clues to a solution. In particular they hold out the prospect of a new theory of settlements which reconnects people to places and also to *presence*, that quality of living in the here-and-now which industrial urbanism has so weakened. One study by environmental psychologists, for example, found urban-rural differences among the same group of people. The Abaluyia are a Bantu sub-clan in Kenya who alternate rural horticulture with city labouring work. In the city, the researchers found, the children were more aggressive, threatening and disruptive. In the country, they were more helpful, co-operative and sociable. The immediate family setting, conventionally nuclear in the city but extended, in the country, beyond parents to other relatives, was critical. In the city children and mothers spent more time together but with fewer routine tasks. Yet the family setting could not be separated from the wider settlement pattern.

The 'pull' of the city is thus, it seems, a kind of force-field determining behaviour in ways that often have to be described as perceptual or illusory. And just as the strength and structure of a magnetic or gravitational field cannot be separated from the bodies causing it, so the urban force field is a precise expression of the way in which a city's component parts – its people and its places – are arranged.

Each city, to that extent, is unique: its pull and image are entirely its own. Predictive models and public policies which do not allow for this can only be partially effective, a factor which probably explains why central government initiatives, compelled for reasons of political expedience and administrative ease to assume a monolithic 'inner city' made up of homogenized places and undifferentiated human masses, have been so unsuccessful. The corollary is that policy-making should be, as it were, miniaturized, to the point at which real landscapes and recognizable groups of people emerge out of the fog. But how does this translate into practice?

The study of the Bantu sub-clan quoted above relies on the notion of the 'behaviour setting', a concept of fundamental importance to the new school of ecological psychology emerging during the 1960s. A behaviour setting is a place but it is also the people in it: the two aspects cannot be treated in isolation. People who enter a setting, ecological psychology declares, 'are pressed to help enact its ✳ programme'. Settings also have optimum 'manning' levels, measures, effectively, of both density and participation. But perhaps the most original feature of ecological psychology, one of great relevance to the greening of the cities, is its immensely fine-grained fieldwork in the community. Neighbourhoods are surveyed in exhaustive detail for a comprehensive account of their behaviour settings.

Studies by ecological psychologists have produced revealing findings. Smaller schools, for example, yield more settings per pupil, thus increasing participation, and drawing in academically marginal pupils. Small-town children have a richer and more finely articulated knowledge of their immediate neighbourhood than city children. They revisit settings more often and use more symbols to describe a smaller area. In a small high school, says one leading proponent, students behave 'small high school'. In a large university they behave 'large university'.

Certain broad themes emerge from ecological psychology and from related research in environmental and developmental psychology. These not only stress the vital role played by 'present space' – let us call it the home ground – in individual and community identity but show how different ways of exploring the home ground affect the identity-building process. In his essay *Psychology and Living Space*, based on surveys of people's mental maps of their home areas, Terence Lee demonstrated a positive relation between local social involvement and the physical pattern of the area, and arrived at a measure which he called the 'neighbourhood quotient'. Involvement took time – ten years of residence, he concluded – but was greater among the families of men who worked locally.

Later studies have shown that walkers or cyclists have more detailed mental maps than motorists: young cyclists are actually more satisfied with their neighbourhood. This is hardly surprising when one considers Piaget's view that knowledge is 'acting-in-space' not 'perception *of* space', that psychological development, in other words, is bound up with active exploration. Children in cities, however, find opportunities for such self-reliant exploration and development in extremely short supply. Parental worries about traffic or about crime make children more dependent on their family, more child-like. Four-fifths of children aged 9 and over owned a bicycle, one study found, but only 2.5 per cent were allowed to use it for school. City children, meanwhile, travel further afield than the small-town child and visit more settings. But although they 'see' more, they 'learn' less. Urban adolescents, Kevin Lynch concluded after his international survey *Growing up in Cities*, were victims of experiential starvation. Broader conclusions can also be drawn. One comprehensive analysis of stress found that the most powerful predictors of psychological health were a coherent world-view coupled with some means of participating in the events around one.

Slowly a picture takes shape, helping to explain why throughout the expansion of cities the Victorians clung desperately to their vanishing villages, and the twentieth century, in its positivist and convoluted manner, has done much the same. Few illusions probably remain about the pre-industrial village. As the historical sociologists like Peter Laslett have shown, it was a hard, tight, stratified world. It was, however, relatively autonomous and self-contained, in many cases self-governing. Land-use was both private and communal. So, in consequence, was work – spinning, child-rearing or garden husbandry at home, and, beyond home, labour in the open fields and the commons and on the home farm of the 'lord of the soil'. It was, indeed, tiny: well over four-fifths of the population in late Stuart times lived in villages containing on average 300 people. It was also, by modern standards, undemocratic.

There are several striking details about this portrait of the pre-industrial village. There is, for example, the parallel between its tininess and the size of the home area or defended neighbourhood spoken of by the environmental psychologists, both only a few hundred people. There is the shared life, existing largely by virtue of shared space – common activity, or community, is physically impossible where there is no common ground. There is, finally, the physical identity and autonomy: the mixture of *genius loci*, local autonomy, and a sense of belonging, each contributing indefinably to the other.

In a culture saturated with positivism, disciplines like environmental psychology provide a publicly acceptable language for confused, inchoate and essentially private emotions. And what these disciplines herald is a widespread desire to rediscover and recolonize the home ground as a prelude to any grander undertaking. In part this is a reaction against ideology. It is also the mark of a self-preoccupied culture experimenting with a wider social world beyond the self. The experiments thus occur in the neighbourhood, one step – the first step – outside the private domain. The village, like the city, is organic and self-willed but its territory is small and distinctive and its operations limited, visible and accessible. What is ultimately at stake, however, is an ideal of communal organization – individuals uniting together for shared ends and wider benefits – which runs counter to the grain of an emerging individualistic culture but which, more importantly, has been discredited by its attachment to the super-organisms of city and nation, with their attendant bureaucracies and vested interests. In effect, it is an attempt at a new definition of community somewhere between a rediscovered private realm and a renounced public realm. Hence it builds carefully, exploratively, piece by piece – as the saying goes, 'from the bottom up'.

The 'new localism' is a sign of this reawakening. It has taken several forms. The remarkable growth in local history and in the urban studies and heritage movements have already been noted. All testify to a resumed quest for roots. An upsurge in 'environmental localism' has also occurred, dating from about 1960 but concentrated in the early 1970s. County naturalists' trusts grew from a membership of 5,000 in 1961 to 155,000 in 1983, while the number of local amenity societies rose from 300 in 1960 to 1,250 in 1975.

Unemployment has meanwhile prompted the emergence of a new local economy, particularly in the cities, stressing small-scale and home-grown employment as a self-conscious reaction to reliance on the 'branch plant' economy of the big national and multinational corporations. Big business itself has become more locally involved: among the best-known examples is Shell's Better Britain Campaign, launched in 1982. The self-managed economy, an emphatically local sector, has also expanded during recession, not only co-operatives and self-employment, but also initiatives such as community businesses and small 'workspace' projects. By 1985 there were 215 local enterprise agencies involved in a sixth of all jobs created.

But the new localism has spread its roots even wider. Both the tenant co-operative movement and the self-build housing movement are attempts to regain control over home space which has been lost to absentee authorities. By 1984, against immense odds, the self-build movement was effectively second in the league of volume house-builders, putting up nearly 10,000 houses a year.

A further approach is to treat the city as an arena for adventure, in the sense usually associated with the distant green landscapes beyond its boundaries. If the inner city represents the opening of a new frontier at the heart of civilized life, then the tangled growth of its waste land is a new wilderness, providing challenge, excitement, a mental and spiritual diet of extraordinary richness. The 'danger-park' – the Ackers in Birmingham, for example, has a thirty-foot artificial rock-face for climbers, the first of its type in the country – is a man-made version of this. But the idea has been taken much further by bodies like the National Association for Outdoor Education, founded in 1970 and headquartered in Doncaster. From here 'cycle safaris' are staged, treks organized around slag heaps, nocturnal canoeing expeditions despatched along canals. In Sunderland the Urban Centre for Outdoor Activities has used reclaimed town-centre industrial land for rock-climbing, orienteering and campcraft. Adventure thus becomes inward bound, reactivating local space and teaching the kind of self-reliance learned in a vastly different setting by Arthur Ransome's *Swallows and Amazons*.

Self-reliance is for the individual what localism is for the community. Both are assumptions of power, declarations of independence. Most important, the self-reliant individual is an indispensable part of the self-reliant community. Both require exhaustive self-analysis and self-discovery, the methods of ecological psychology. To express it more conventionally, the community has to find or make its own reserves and resources. It has to create itself and thereafter to renew itself continually. Its resources are a place, its people, their talents and time: from these shared ends will flow. This, then, is the new working definition of community, a small, fragile, unique organism, a creature of a particular time and place, which cannot be implanted or cloned and which can never be taken for granted. It is, in consequence, unamenable to fiat, edict and regulation. It cannot be legislated or hypothesized into existence.

The self-reliant individual in the self-reliant community forms the theme of many initiatives since the 1970s. He figures in nationwide award schemes like the Royal Society of Arts' Education for Capability. He figures in the urban studies and environmental education movements. He figures, most importantly, in an approach which offers demoralized communities the skills they need to rebuild themselves. The Skeffington report in 1969 said the practitioners of this new approach should be called community development officers. Grander names suggested include animators, enablers and facilitators. Thanks to the work of Rod Hackney in a run-down terrace of houses in Macclesfield in 1972 and its endorsement by the Prince of Wales in 1984, that branch of the business known as community architecture has entered the realm of public discourse, but there are also, of at least equal importance, community planners and community landscape architects. In the United States, where urban renewal has been seen increasingly as renewing people as much as places, they are called community organizers, who work with neighbourhood development corporations and receive training from at least nine specialist national centres.

Alongside this new breed of person has grown up a new breed of organization, antiseptically labelled the community technical aid centre. Typically this consists of a group of professionals serving a particular neighbourhood, living and working in it, supplying skills to local groups who may want to turn an old church into a community centre or to landscape derelict land. Its origins lie partly in the community design centres of the 1960s.

By 1983, when the Association of Community Technical Aid Centres (ACTAC) was formed as a national umbrella group, nearly 100 agencies were providing environmental design aid for local voluntary groups. Demand for their services was overwhelming. In 1985 the cause of local self-reliance received an enormous if belated boost with the establishment of the first national co-ordinating body, the National Structure for Community Development.

Aid and animation, however, are ultimately no substitute for community self-help, and it is here that the most significant developments have come. Environmental psychology turned cities upside-down, eyeing them from street level rather than from cathedral parapet or penthouse suite. Ecological psychology delved deeper and broader, piecing together individual perceptions into shared settings and relationships. Ecological landscape design, expressed most powerfully by Ian McHarg, showed how the ancient principle of *genius loci* could supply a method by which communities could collectively map their own distinctiveness, in soil and water and landscape types. 'Place,' McHarg wrote, 'is a sum of natural processes and … these processes constitute social values.' And finally, in the work of Christopher Alexander, do-it-yourself becomes a design philosophy capable of embracing anything from nailing a window to creating a city-region. In Alexander's *A Pattern Language*, published in 1977, the business of design is painstakingly unpicked and demystified. It is, in effect, a manual for 'architecture without architects'. Alexander's premise is that all those who share an environment, whether a house or a city, should co-operate in designing it.

In that curious fashion ideas have of sprouting up hybridized and camouflaged in unexpected places – what social historians used to call the *zeitgeist* or spirit of the time – the years since 1970 have seen this broadly based ecological approach permeate many different initiatives aimed at rescuing failing communities. In the summer of 1983, for example, in the run-down Conway district of Birkenhead, residents exhaustively surveyed the natural resources – land, buildings, equipment – and the human talent lying untapped around them. 'What are you good at?' was the question asked on doorsteps. Tony Gibson, whose Nottingham-based Education for Neighbourhood Change programme provided the back-up, calls the process 'community double-digging', and it serves as the prelude to 'planning for real' in which people assemble a giant cardboard model of their neighbourhood so that, collectively, they can decide its future shape. Gibson has developed an armoury of community building kits which serve to formalize what in a healthy community is instinctive. The similarity in method to the ecological psychologists is striking.

Running through this and other attempts to rebuild communities, like a theme with an infinite number of variations, is an insistent preoccupation with the land and its products, with the countryside and its imitations, with the natural world, its cycles and life and laws. One calls it green because that term, and only that term, adequately and invariably defines the symbolism people use to express their needs. But for all the moral and imaginative force of environmentalism, one does not need to invoke eco-spirituality to account for it. There are many simple and practical reasons.

Greening is a therapy, a technology, an art, a science and a way of recovering a lost identity. For the trapped of the cities, it represents the most direct path from halfness to wholeness, since it was their 'countryside' self that the city stole from them. And the neighbourhood nature park or wildlife garden proclaims the same of the community, reminds and reassures its members daily of their new enhanced identity.

That it is both a therapy and a technology should surprise nobody. The garden and the countryside are among the most important sources of our creative leisure: recreation surveys repeatedly attest this. They evidently supply some of our deepest psychological needs, so much so, in fact, that various forms of clinical treatment have been built around them. Wilderness therapy, for example, is practised widely and with impressive results by psychiatric hospitals and juvenile centres in the United States. Horticultural therapy, a feature of many city farms, has achieved similar successes with offenders and prison inmates and with the mentally disturbed and handicapped in both America and Britain.

The gains of the latter have been carefully evaluated. The picture that emerges is of a co-operative, accessible and creative activity providing a 'touchstone with reality' in the form of natural cycles. It sublimates aggressive impulses and gives scope for self-determination and 'personal space'. One survey of gardeners found their main source of satisfaction, cited by 60 per cent, to be 'peacefulness and tranquillity'. Plants delight and respond: they do not judge, order, threaten or demand.

For apathetic people and broken communities, such therapy becomes an accessible, that is, an appropriate, technology. With it comes self-belief and with self-belief comes something of the spirit of the entrepreneur. For communities the powerful imagery is that of land-work, visible and dramatic change wrought by the simplest of tools yet of a piece, ultimately, with that most sophisticated of sciences, design with nature. This does much to explain the popularity of environmental work in job creation and urban aid, and shows why the attempt to make cities productive, of food, energy and raw materials, may well be the most promising route for their recovery. It also explains why many groups launched simply, often naively, to make cities more environmentally tolerable have ended in the creation of jobs in fields far removed from landscaping: in Bristol they have ranged from handcrafts to computers. These are the 'ripple' effects of returning self-confidence. They demonstrate why entrepreneurialism is a social as much as an economic resource. They also help to define the new style of neighbourhood technology.

Neighbourhood technology often operates by accident. When a recycling network or a city farm or a loft insulation service is set up, people meet, talk, forge bonds, act jointly. When some piece of forgotten countryside is defended against building or municipaliz- ation, isolated individuals form campaign groups which in turn grow into networks of friends, wardens, even managers: a process as true of Moseley Bog, Highgate Cemetery and countless other secret green places in the 1970s as it was when the Hampstead Heath Protection Society was formed eighty years before. When individuals share ideas, tools and cups of tea in the making of a nature garden from some rubble-strewn eyesore, they are creating not only the rudimentary ties of community but the shared space all communities need in order to celebrate and renew themselves. Social psychology tells us that such neutral areas are required to maximize interaction. But in effect they recreate the common land of the hand-made village, long since lost to the absentee landlords of government and industry. For comparison, look for the shared space, the symbols of commonalty, in the contemporary city and one's eye lights, ultimately, on some distant town hall and its empire of 'services' and 'facilities', run by a fully professionalized elite grown too expert in the arts of spoon-feeding.

In the new village common of city farm and nature garden, planning for real becomes collective landscape design, a branch of the art wholly distinctive to the greening of the cities. *Genius loci* has been democratized, more truly a spirit of place because it is also a spirit of people and a place cannot be separated from its people. Through that place, and others like it, the city is being co-operatively redesigned: its waste space thus offers scope for community creativity, for a whole neighbourhood to display its talent, its industry and its sense of identity, probably without precedent. Searching for parallels, one thinks of the medieval church, where common enterprise reached towards a shared and highly visible symbolism.

These are some of the reasons why in the priority estates programme at Tulse Hill, south London, children built a nature garden, why the Weller Street housing co-operative in Toxteth, Liverpool, wanted trees and flowers, and why the self-surveyors and fact-finders of Birkenhead chose gardening as the cement in their self-help fabric. A garden shop, club and display area were planned in a disused art college. From this new neighbourhood centre, a gardening service would also be run. Altogether 822 gardens were counted in their door-to-door enquiries. 'Gardening interest which begins at home,' they concluded, 'often extends to a general concern about local streets and open spaces.'

The recovery of greenspace also serves to define the urban village as it surfaces after its long submersion in the city. The thrust of research in environmental psychology, for instance, is that greenspace is a highly effective boundary marker for neighbourhoods, lending distinctiveness and thus cohesion. The important lesson remains, however, that communities define and create themselves. Neighbourhood planning – a sort of packaged togetherness – is in this respect as much a symptom of industrial society as a cure.

The greening of the cities has demonstrated conclusively that people love those places best where they feel they most belong. By now, perhaps, this seems an unexceptionable conclusion but its implications are explosive. Take an act as simple as planting a tree. Consider two versions of it. First there approaches the council manual employee, unionized and piece-working, aware only of another job to be done, another hole in another pavement in another district. He plants and cages the tree, for it has to be protected against the natives, and in due course it is defaced or uprooted. In further due course, after allowances for committee authorization procedures and interdisciplinary anti-vandalism conferences, the operation may be repeated. Equally, since it is a relatively expensive operation, it may be abandoned. In either case the tree dies and a small part of the city with it. In time these small parts grow larger.

Consider, now, an alternative. A child plants the tree. Aged, say, in his early teens, he is a routine defacer of trees in remoter parts of the city. But this tree, and this hole, are special. The hole is special because it is in the pavement outside his home and the tree is special because it is a lesson in growth and an exercise in responsibility and an experiment with unknown powers. Most of all, however, it is special because it is *his* tree and *he* has planted it. The tree acquires a guardian: it has less need of the guard.

These caricatures convey a truth. The act – digging, fertilizing, watering in – is virtually identical in its mechanics but utterly dissimilar in everything else. The moment of planting is an infinitesimal fraction of a tree's life: where is the workman for the rest of it? Industrial society has done much to disjoin the mechanical from the human content of technology but in a world increasingly fashioned by and through technology, to ignore the human content is to ignore what is overwhelmingly the largest part, the part represented by the hundred years and more of a tree's life after the thirty minutes taken to plant it.

In economic terms alone the cost of ignoring the human content of technology is enormous. At least three-quarters of the shrubs and standard trees planted conventionally do not survive. By 1981 in Liverpool areas were being landscaped for the third time in a decade. Experience with 'blocklings', tree seeds raised in a three-inch cube of peat at a cost of two pence each, has shown that they are ideally suited to school tree nurseries. Junior schools, it has been suggested, could with a little imagination become the 'largest single agency for amenity tree planting in Britain'. Ultimately, however, far more than the planting of trees is at stake.

In 1969 the American planner Sherry Arnstein invented her ladder of citizen participation. Each rung on the ladder represented a different state in the ascent from a powerless and manipulated citizenry through the 'tokenism' of consultation and placation to partnership, delegated power and, finally, citizen control. Each rung scaled also involves the displacement of some powerful entrenched interest.

Environmental action in the cities has involved the scaling of many rungs. In part it has been a resumed struggle for power, galvanized by unemployment and fought out, as many such struggles have been and as the theoreticians argue they should be, over land. Battlelines remain confused but two sets of protagonists are discernible: on the one side, the consumers, clients, customers and employees seeking to translate the rhetoric of participatory democracy into practice; and on the other, the councils, trade unions, corporations and bureaucracies seeking to stop them, usually 'for their own good'.

Much hard and bitter experience attests this conflict, from the union-inspired ban on voluntary labour in so many parks – once, it should be remembered, called 'people's parks' – to the thirty-seven different council officers who required contacting before a one-and-a-half-acre children's garden could be set up in Stockwell, south London. Max Nicholson, chairman of the Ecological Parks Trust, has spoken of the 'army of technical obstacles' confronting even the most modest project. The social entrepreneur, faced with such malign inertia, requires vast reserves of patience, determination and optimism, a sad footnote to the history of state welfare. What is extraordinary is that anything is achieved in such conditions.

Belatedly, however, changes have occurred. By 1985, one report was able to list over 300 public involvement schemes in eighty-eight local authorities, ranging from regular consultations with residents' and tenants' groups to more adventurous initiatives like neighbourhood planning in Birmingham and the decentralization of offices, in Walsall and elsewhere. Birmingham, for example, has proposed setting up eighty 'urban parishes'. But probably the most significant advances have come in the management of greenspace by local trusts, community groups and user committees. By 1986, for example, the London Wildlife Trust was running twenty-five sites in the capital. Local naturalists' trusts alone managed 100 urban reserves, although in the country as a whole they were responsible for two-thirds of all nature reserves, a total of 1,600 sites covering 120,000 acres.

The cumulative effect of these developments has been to create the outlines of an alternative power network at grass-roots level, a kind of rival local government. It is a 'green' network because it deals in the raw materials of urban redesign – land and landscaping – and because it expresses principles and needs central to deep ecology, most of which have been consistently ignored by local authority and state bureaucracies. The organizations of which it consists are, in general, small, accessible, fundamentally voluntarist, non-hierarchical and project-oriented. They are professional without, as yet being professionalized. They give advice, not orders. They stress participation, not paternalism. They are expert yet they are also, in an older, more dignified sense, amateur. They thus offer a role for the true amateur, the interested and concerned citizen.

This rival pattern of democracy has arisen in response to the challenge of the cities and their failing institutions. The landscapes it has created are outward signs of a widespread, genuine but still tentative social and human development. To a society conditioned to sweeping collective solutions, it may often appear a small patchwork affair when set against the massive scale of urban decay. One is tempted to say it is all there is.

Bureaucratization, of both people and landscape, remains a threat, however. Art, particularly landscaping and architecture, does not operate in a social vacuum. The tower-blocks that litter our cities are as much the consequence of the mass packaging of instant solutions as are soap powders and polyunsaturated margarines. The English Tourist Board has itself stressed that the most successful regeneration schemes are those born of individual commitment. Just as, in the eighteenth century, privately executed landscaping degenerated into whim and proprietoriality, so today bureaucracies tend to produce bureaucratic landscapes. The answer thus lies as much in social reform as in some ostensibly self-contained stylistic innovation. If a landscape, any landscape, is not to grow tired and frayed, it must continue to reflect the needs and loves and imaginings of its users. It must also reflect their actions and decisions. That is perhaps the central message of the greening of the cities.

On the steep slopes of the Gleadless Valley, three miles from the centre of Sheffield, is the forty-acre Rollestone Wood. Through it runs the Meers Brook and within half a mile of it, in a slum clearance estate of the late 1950s, live 9,000 people. Gleadless, like Sheffield itself, is sharply defined physically, permeated and surrounded by green space and woodland, and possessing a strong community identity. Eighty-five per cent of residents want to stay in the area. In 1981 researchers tried to find out what Rollestone Wood meant to local people.

The results, in a strangely backhanded way, were highly illuminating. Almost half the primary schoolchildren used the wood regularly for play – for hiding, climbing, jumping streams and swinging on ropes. By comparison, they said, the park was boring. And those children, not surprisingly, were far quicker and keener in the classroom on anything to do with woods than were schoolchildren further away. Only a small minority – 8 per cent – of local adults visited the wood regularly, although 31 per cent went there occasionally and another 26 per cent had visited it when they were younger. But it was considered of personal importance by 72 per cent and of importance to the area by 86 per cent. 'We can remember the area as farms,' they said. 'We've grown up with the wood. We played in it. Now our grandchildren play in it. It's the only link with the past.'

The value of Rollestone Wood, the researchers concluded, was not that it was a facility or a service or a provided benefit, but that it was permanent, omnipresent, always visible. It was, in other words, just *there*, had always been there and would, with luck, always be there, unimproved, untampered with, itself alone and wholly other. When the message of Rollestone Wood has spread throughout the cities, one may well conclude, their greening will be complete.

13

BEYOND THE CITY

I wish to have rural strength and religion for my children and I wish city facility and polish. I find with chagrin that I cannot have both.

(From Ralph Waldo Emerson's journal)

The one outstanding characteristic of these people as I knew them, and which distinguished them from us, was that wherever they went, they felt they were known .... To them, everything was family .... The Bushman appeared to belong to my native land as no other human being has ever belonged.

(Laurens van der Post, on the Bushmen of the Kalahari desert)

Welwyn is an excellent place to live, he says, though a bit short on pubs.

(Welwyn Garden City taxi-driver quoted in 'Shredded Wheat Country', New Society, *15 March 1984)*

In 1984 the famine relief charity Oxfam caused a minor controversy when it turned its attention to the troubled British city. Two small-scale projects, one in the sprawling Craigmillar estate in Edinburgh and the second in Manchester, received £5,000 pump-priming aid. Oxfam's diagnosis was that developed and underdeveloped parts of the globe had many social problems in common. 'One notices,' said an official, speaking of the two British projects, 'remarkable convergences with the Third World.' The Craigmillar project, where art and woodwork were taught to housewives, was likened to an Oxfam self-help scheme in Zimbabwe.

In the same year the United Nations Environment Programme delivered a deeply pessimistic verdict on its six-year campaign to halt soil erosion and desertification, launched in 1977 in the wake of the Sahel drought. In the few instances of success, it reported, 'ingredients . . . appear to be the small scale of the projects, their relevance to perceived community needs, local field direction and community involvement . . . and ability to learn from mistakes.'

It could equally well have been a verdict on the British inner city. On both fronts derelict land has proved a conspicuous symbol of communal failure. Responding to this failure has involved reappraisal of a society's deepest-rooted social and political relationships. In the cities, this has produced new kinds of organization and new philosophies on the relationship between the individual and society which are transforming the vocabulary of political debate. Most important of all, however, is the perception, whatever appearances may suggest, of an underlying convergence between the development paths of global North and South, between what might be termed the post-industrial and the pre-industrial worlds.

The issue of city and countryside is central to the debate about aid, development and global survival. On one hand is the 'growth-pole' model of development, envisaging the Third World undergoing a vastly exaggerated replay of western urbanization and industrializ-ation, with population increase slowing, through the demographic transition, and economic growth 'trickling down' to the rural areas. The growth-pole model, hitherto dominant, is essentially urban and centralist: development is controlled from the top down. On the other hand is the emerging alternative of 'bottom-up' grass-roots development. Decentralist and rurally based, this is founded on appropriate technological aid, technology which fits easily and intimately into lives and communities.

The growth-pole model has proved of dubious efficiency both in diffusing growth and limiting population. It also suffers from one overwhelming defect: it threatens enormous, potentially crippling, environmental stresses. Ecological stress caused by human impact has already meant that six times as many people died in 'natural' disasters like floods or drought in the 1970s as in the 1960s. But the Third World megacity, growing at breakneck speed, is far more than a poisoner and a polluter on an unprecedented scale. In part because of the gross urban bias inherent in the growth-pole philosophy, it acts as a giant distorting lens through which western appetites, and their imitators, are focused on a parched and eroding landscape: an urban Indian, for example, uses twenty-eight times more electricity than a villager. A world fully urbanized on current western lines, its population stabilized in the twenty-second century at eleven billion, would require 255,000 square miles of road, an area three times the size of Great Britain, simply to drive its cars along.

Yet if the history of development has proved anything, it is that aid is a contagion. Attitudes and technologies cannot be sterilized at border crossings. For the west to promote the village in the Third World while meticulously preserving its own air-conditioned, top-down urbanism is thus both a hypocrisy and an impossibility.

Hence the fundamental importance of developmental convergence, of global North and South elaborating, through popular action rather than government proclamation, both a common vision of the future and a means of achieving it. And convergence is a growing practical reality, producing, in the two settings of western city and Third World village, ostensibly at opposite ends of the development spectrum, remarkably similar answers. The British group Green Deserts, for example, promotes tree-planting in the Sudan and urban forestry in the UK. Community forests planted in the Himalayas show tree survival rates of 80 to 90 per cent, at least twice that of state government plantations. In Africa the work of the Kenyan Green Belt movement involves local groups – women, schoolchildren, small farmers – planting woodlands, for fodder and fuel and to halt erosion. By 1985 16,000 had been planted and the movement had spread to Tanzania, Somalia, Sudan and Ghana.

The promise of convergence is immense. If North and South can map out a shared development path based on a new ecological relationship with the earth, its cycles and rhythms, many difficulties of aid and trade will vanish and many lives and landscapes will be saved. The greening of the cities showed the direction such a path might take. Founded on a wealth of small-scale initiatives, it sought, above all, new land, land in some cases physically reclaimed from corrosion and blight, in many other cases mentally reclaimed from the quarantine, the emotional anaesthesia, engendered by industrial culture. From this land, new powers, products and human prospects would flow; on it and through it, community and creativity were rediscovered, a fresh vision of nature explored and celebrated, bonds between people and place renewed and strengthened. From the land also, albeit haltingly, came global awareness, the outlines of an ecological ethic, planetary in scope, revolutionary in its implications, and immensely subversive of habits of behaviour and patterns of settlement of which western industrial urbanism was the culminating expression. The new emphasis was on the local, the dispersed, the home-grown and near-at-hand – all of them, in one sense, aspects of human reintegration and self-government – and although there were powerful countervailing forces, the vigour and impact of this mood was sufficient for it to be identified and assimilated by the established agencies of state and industry, where it has increasingly coloured an attempted programme of urban renewal.

The keynote of this programme was thus environmental quality. Yet a symphony orchestra cannot play a string quartet. State-backed urban renewal, although it has proved responsive to the new mood, cannot emulate its achievements without a radical change in its own constitution and obligations. Chief among the latter is the strange cultural imperative encountered in the earlier chapters: the desire to perpetuate the city, and a corresponding conviction of its perman-ence, in the face of mounting evidence to the contrary. The grass-roots urban greeners sought to transcend that imperative and to unmake the city. Until the dimension they brought to the redesign of cities is fully acknowledged and accepted, conventional regeneration is merely the plugging of fingers into a permanently leaking dyke.

The burden of this argument is that, over the long term probably even more than the immediate future, the flight from the city will continue, dereliction will grow and investment in the urban fabric will in an important sense be wasted. The economic and industrial arguments for centralization have, with a few exceptions – the importance of 'face-to-face' contacts at senior level, for example – largely vanished. In both a spiritual and a political sense, power is no longer concentrating in one place or one body but is dispersing outwards and diffusing downwards. But there are even more fundamental factors at work, centring on the peculiar cultural and psychological fixations that have formed such a recurrent theme of the last dozen chapters.

The countryside has come to represent a primary value for western society. It is the chief arena, outside the home, of our leisure and pleasure, the chief target of our socio-economic aspirations, the touchstone, increasingly, farmers notwithstanding, of our notions of aesthetic beauty, psychic freedom and, perhaps most crucially, spiritual reality. Walling ourselves off from it in vast conurbations merely serves to intensify the yearning.

The city attempts to compete with this. It improves environmental quality, bans pollutants and creates surrogate countryside within its boundaries. Yet those boundaries remain. And it is precisely because of those boundaries that city and country remain separate places and that people escape from one into the other: a movement almost gravitational in its strength.

This was a reality recognized by Ebenezer Howard and it is implicit in the spread of greenways as a principle of urban design. Its message is that the city must be unmade by the unmaking of its boundaries. Profound cultural and psychological forces are involved. That these are intimately bound up with our sense of the sacred is shown by the 'omphalos' legends of early cultures in which chthonic serpent or dragon powers, hitherto untamed and free-ranging over the earth's surface, were bound and harnessed to a particular centre: the omphalos or navel. In Greek and Roman foundations of a city, the plough which delineated the original boundaries, creating the *sulcus primigenius*, or primary furrow, was lifted clear of the earth over areas intended as gateways, leaving the free flow of the spirits – the ancient dragon powers – uninterrupted. Hence the role of greenways in breaking up our sense – our 'image' – of the city. Against the background of a recovered primitive, the new forms of earth-worship or immanence, for example, greenways become spirit-paths or dragon-roads. They express spiritual freedom, reconnection with a deeper mode of being. And since it is this sense-image that ultimately dictates how we physically shape the world, the rise of global awareness in human beings, reinforced by the picture in mental space of a round, finite, threatened and beautiful earth, is of great importance. For the planetary city – its factories in the Third World, its banks in the west, its parks in the rain forest or the ice-cap – simply renders redundant the lesser division between city and countryside. Settlement, or more precisely, the *sense* of settlement, has become ubiquitous. The city is a vestigial organ, a vast fossil-outline washed up on the beach of history, stranded somewhere between a resurgent localism and an advancing tide of globalism.

Ultimately, therefore, cities will merge into countryside. In the older industrial city, the process must be hastened by the creation or re-creation of green space on a massive scale and the colonization and management of that green space to achieve some mixture of farmed countryside, woodlands, water, villages and non-polluting industry. A corollary is that the population outflows are accelerated, with many millions more moving out into the countryside: the resuscitation, in effect, of something like the New Towns programme.

Decay in the old cities should be speeded up by demolition. The historical trend towards lower-density living should meanwhile be pushed further, to the point where small-scale farming and related occupations became a possibility for thousands, perhaps millions, more people who may move into the countryside. In both cities and countryside, the relationships between men, plants and wild creatures will grow richer, more subtle and various, so that old distinctions – wild and tame, productive and ornamental – became blurred and redundant.

As a further corollary, land now imprisoned within these categories, land from which settlement is barred (green belts) or production is barred (sites of special scientific interest) or wild creatures are barred (parks and farms), will be freed. Human beings will begin to treat the countryside as home rather than as a hierarchy of no-go areas barricaded in behind legislation, subsidies and planning restraints. And perhaps most crucially, and posing the greatest challenge to social ingenuity, this vast resettlement and re-education programme must be planned, managed and organized so that it is *not* planned, managed and organized. The communities of tomorrow cannot be coerced into existence. The state must contrive to make its operations invisible.

Curiously enough, only a minimal change of emphasis in policy is required to achieve much of this. Chunks of cities are already being demolished, but through errors or neglect. Many old roads, houses and factories already lie beneath new urban countryside. At Windmill Hill, in Bristol, in a way that would have delighted Richard Jefferies, a street was first closed and then buried under orchards and gardening plots. It now emerges briefly at one point to form the farm track. Digging up cities deliberately and in a transparently honourable cause is infinitely preferable to doing it by default.

It is also justifiable on straightforward economic grounds. Lack of space in cities is a besetting obstacle to job creation. Building up a land-bank in the shape of reclaimed countryside is thus a long-term strategy for attracting that industry back. By the same token instantly plugging up holes in the city with new houses, under the slogan of 'homes with gardens for local people', may well amount to a peculiarly refined form of social torture, since jobs will continue to leave the cities and the householders will be immobilized into a state of permanent unemployment and economic dependency.

The issues are urgent. In jumbled and inchoate form, the fragments of a new land-use settlement for the post-industrial age lie around us: renewed housing pressures; mass unemployment coupled with a slow revolution in working patterns; an upheaval in farming; the unprecedented freeing of space in cities; the achievement of national population stability; and the radical revaluation of attitudes to countryside, wildlife and food production that are the most obvious fruits of environmentalism. History only needs a small nudge for these elements to fall neatly into place and for the outlines of a post-industrial land-use settlement to become clear. A closer look at each of these elements will explain why.

First, substantial changes in the planning of cities and countryside are an inevitable product of housing need. The inexorable thrust towards more spacious living is a constant of the modern historical record, attested equally in preference surveys and land-use data. Between 1901 and 1971, even through the era of tower-blocks, urban densities continued to fall, in total by about two-thirds. By 2001 densities will be half their level of a century earlier.

Survey evidence, meanwhile, monotonously reiterates the message that the pursuit of more space is probably the chief reason, job changes apart, why people decide to move house. A better neighbourhood – the physical and social characteristics of an area – is an important secondary aim. Evaluation of residents' reactions to housing produces other interesting findings. Most people's ideal dwelling is a detached or semi-detached house with a garden. Far more weight is attached to the landscape of a house than to its architectural qualities: 'homely' materials are preferred. From their windows, people like to see an open prospect, with nearby greenery and some human activity: they do not like to see blank walls, parked vehicles, other buildings too close. One particularly original piece of research used people's reactions to photographs to produce two key factors governing housing preferences: the first linked with the physical quality of the house, the second an environmental quality associated with greenery, privacy and open space, and labelled, provisionally, 'harmony with nature'.

The smaller a settlement in population terms, the more generous is its land provision. Cities of over 500,000 are four times denser than villages of under 10,000. This is known as the 'density-size rule'. At the same time much speculation has centred on the notion of a 'pivotal' density, a norm towards which both sparsely populated and crowded districts are moving. In land-use terms, this may represent a deep-seated movement towards greater equity, a process also reflected in the blurring of city and countryside. Demand for gardens, for which surveys indicate a high income-elasticity, is undoubtedly one facet of this.

Given all these factors, however, the residential momentum outwards from cities quite clearly cannot be halted without radical changes in city form. Moreover, it is clear that the potential contribution of cities to the estimated annual requirement for new houses in England and Wales until the end of the century – between 130,000 and 250,000 – is marginal: an absolute maximum of 20 per cent and probably nearer 10 per cent. To achieve the higher figure would almost certainly involve the construction of a new generation of cramped, low-rise slums, destined to become rapidly obsolete, in callous disregard of the quest for urban space and 'harmony with nature'. The greenspace which is at long last beginning to make living in many parts of our cities tolerable would be obliterated. The city's long-term prospects of revival would thus be destroyed without – and this is the true irony – any serious reduction in conflict over housing in the countryside. Hundreds of thousands, meanwhile, would remain incarcerated in tower-blocks. These are the implications of shire-based calls for urban regeneration.

A second major element in the emerging pattern of post-industrial land-use is the crisis in farming, which has created a window of opportunity for a policy of more broadly based rural resettlement. Modern farming's chief efficiency is its employment of fewer and fewer people in the cause of labour productivity: a dubious virtue in an age of apparently permanent mass unemployment. Its chief inefficiencies are its wasteful use of land and energy. The point was registered almost two centuries ago by William Cobbett – 'The cottagers,' he remarked, 'produce from their little bits, in food for themselves and in things to be sold at market, more than any neighbouring farm of 200 acres' – and it is no less true today. The smallholder's detailed and labour-intensive relationship with his plot produces far higher yields per acre for a far lower commercial energy input. The traditional Chinese peasant practising intensive polyculture is thus 110 to 120 times more 'efficient' than the modern British agribusinessman.

Small farming, in other words, saves land – and so too does the fall in meat and fat consumption, and the switch to a diet based on fibre, fruit and vegetables, which probably owes as much to increased vegetarianism, another product of the environmental revolution, as to the new awareness of diet and health. The average Briton needs an estimated 1.6 acres to feed him. An all-vegetable diet would cut this by almost two-thirds to 0.6 acres. At a strategic level the consequences are startling. A meatless diet would free thirty-six million acres from farming, roughly four-fifths of the total agricultural area and three-fifths of the UK's land surface. Only 2 per cent of the land now farmed would be needed if intensive gardening for food was practised. In Russia, allotments produce 84 per cent of the food on 4 per cent of the cultivated land.

In 1985, the year after the National Farmers' Union own long-overdue review of farming practices was launched – the NFU called it Agriculture's Watershed – influential figures within the industry were at last beginning to confront the consequences of redundant farmland. Estimates ranged from 1.7 to ten million acres. The ostensible cause was a vast food surplus, ascribed to the EEC's agricultural policy and to decades of extravagantly subsidized farming malpractice. Both the NFU and the Country Landowners' Association indulged in strange new talk: of chemical quotas, of part-time farmers – numbers had already increased by 7 per cent in the previous two years – of 'a wide range of small-scale commerical and industrial ventures'.

The expensively won wisdom of the farmers' leaders bears a curious resemblance to the new community of Lightmoor in Shropshire, the 'garden city of the twenty-first century' intended as a reinvention of the ideas of Ebenezer Howard. Here, on twenty-three acres of land to the south of Telford new town, twelve households will combine part-time or freelance 'outside' jobs – teaching, bricklaying, computer programming – with home-based horticulture, husbandry and rural crafts, and also with backyard high-technology industry: computer software, for example. Lightmoor will be co-operatively run, designed and built by its residents, its housing incorporating energy-saving and recycling technology. Ultimately, it will form one of several electronic hamlets arranged in clusters on a 250-acre site.

Lightmoor, a collection of level-headed people dissatisfied with urban life, may serve as a symbol of a potential rural renaissance, a distillation of thousands of unco-ordinated and apparently unconnected initiatives and aspirations. Behind it, for example, lie the dozens of communes, co-operatives and radical social ventures which, particularly since the 1970s, have sought to revive communal identity through a new relationship with nature. Many perished, but a surprising number remain and prosper, united in Britain, for example, by the Alternative Communities Movement and the Communes Network, earning a living from farming, visitors, light industry and crafts. They range from avowedly New Age bodies like Findhorn in northern Scotland, where animism has sprung new roots and found a new international audience, through looser networks of ecologically conscious crofters and peasants like Scoraig or Tintern, to more socially committed self-planning groups like Greentown in Milton Keynes.

Behind Lightmoor, too, lie experiments like integrated rural development: the attempt first in the Western Isles of Scotland and later in Derbyshire to re-energize upland settlements by introducing environmental improvements, better community facilities and more business opportunities – by reducing, in other words, the human monoculture involved in economic dependence on a failing agriculture. And behind it, more broadly, lies the great mass of movement out from the cities of people for whom remoteness is no longer a barrier and who clearly share many of the deeper impulses and aspirations of the Lightmoor pioneers.

Those aspirations are reflected increasingly in the shaping of new cities, particularly since the 1970s in the places where the New Towns programme has been infected with the spirit of ecological radicalism, at Redditch and Telford, for example, and, more innovatively, at Warrington and Milton Keynes. Yet many less conspicuous free-market initiatives display the same characteristics: not merely the search for 'harmony with nature' but the quest for local cohesion, involvement and self-determination. In Kent, for instance, there is the clustered village of New Ash Green, a speculative green belt development where old woodlands were combined with extensive landscaped areas to provide 'a thread of the Kent countryside running through the village'. At New Ash Green the community itself, through the village association and residents' societies, took over management of its own landscape.

Probably the most radical innovations, however, have come in Warrington. Here, in districts like Birchwood and Oakwood, cells of housing and industry are set in, and separated by, a landscape web: a tangled mix of wild flowers, shrubberies and woodland which sweeps up to garden gates, so that residents step straight out into a green country lane. Sites for reclamation and landscaping are treated as unique ecological and historical records, each one a compendium of small identities. Even a wheel rut or a rabbit scraping, say Warrington's landscapers, can vary landform and, with it, plant life. Land engineering is based on natural principles: pond edges are scalloped, ditches take the place of drains. A favourite maxim is: think what a farmer would do – and do the opposite. The landscape created is cheap, dense, resilient and highly attractive. Birchwood Science Park, set in its midst, has proved extremely successful in attracting jobs.

But the Warrington approach goes further. It attempts to eliminate the schisms between designer, manager, producer and consumer. It also involves landscape 'after-care' – equivalent, say Warrington planners, to taking out a maintenance contract on people's minds.

In 1977 an environmental education team was established in the New Town, followed by a professional park ranger service in 1979. In 1981 teams of voluntary park rangers began to be formed. Ponds are managed by anglers' groups. Local children become 'ranger helpers', patrolling and maintaining the parks, planting trees and wild flowers, helping out at events: 55 per cent of Oakwood's children have joined the helpers' club. Parks are used for work, play and study, for carnivals, bonfires, guided walks, natural history workshops, nature games, 'oak-hugs', weather recording, and building bird boxes and tables. Moreover parks are well used. In 1984 215 events were organized in the five main park systems, attracting 37,000 people. Tree-planting by primary school children is almost an industry. An entire school year often spends a whole day in the park. Dozens of school grounds, in return, have been planted as nature reserves. The rangers, active in schools, and immensely popular with the children, take part in community development meetings, an explicit recognition of the wider social role of environmental awareness.

By 'marketing' an environmental ethic, the Warrington rangers have practised preventive social medicine, protecting a vast financial investment in roads, buildings and landscape. There is an 'almost complete lack of vandalism' in the parks, which can thus afford to provide a wider mix of features. But the real lessons go deeper, beyond social engineering, however discreet, to individual relationships. People cannot care for land, Warrington shows, without land, in some sense their *own* land, to care for.

Milton Keynes, as befits the design of a city rather than a town, offers clues on strategy as well as tactics. There is, for instance, its concept of the urban forest. There is the return to the lower densities of the earliest New Towns, complemented by a polycentric settlement pattern based on identifiable neighbourhoods. There is the extension of the greenways concept into an ambitious linear park and open space strategy using river valleys, lakes and the Grand Union Canal as a frame on which the city can be draped. There is the attempt to interpret local identity ecologically: the division of the 'string-bag' road system into six 'species zones', each planted with dominant species of tree and shrub.

There is also the sheer lavishness of the parkway landscape – two twenty-four feet carriageways set in a woodland belt up to 330 feet wide – and its complementary 'redway' network of cycleways and footpaths. At Milton Keynes cycling and walking have become technologies appropriate to a city. Cycle ownership is two-thirds higher than the national average, while accident figures for cyclists and pedestrians are, respectively, a half and a quarter of national rates.

Milton Keynes's endorsement of both self-build and solar technology are particularly significant, the former reflecting an awareness that the best communities create themselves by organic growth. By 1984 5 per cent of its private housing was owner-built, on

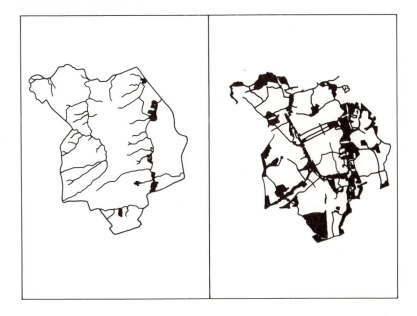

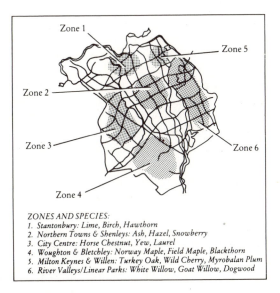

ZONES AND SPECIES:
1. *Stantonbury: Lime, Birch, Hawthorn*
2. *Northern Towns & Shenleys: Ash, Hazel, Snowberry*
3. *City Centre: Horse Chestnut, Yew, Laurel*
4. *Woughton & Bletchley: Norway Maple, Field Maple, Blackthorn*
5. *Milton Keynes & Willen: Turkey Oak, Wild Cherry, Myrobalan Plum*
6. *River Valleys/Linear Parks: White Willow, Goat Willow, Dogwood*

Figure 13.1   Geomorphism at work: Milton Keynes's linear park system, based on river valleys

plots of up to half an acre. And in the 200-acre energy park, begun in 1985 and destined eventually to house 3,100 people, wind, photovoltaics, insulation and, most innovatively, the shaping of land and buildings to improve microclimate, are applied to the design of houses, offices, factories and community centres.

Using solar technology as the basis of a new settlement reflects an even more fundamental principle: design with nature. Energy, as we saw earlier, now constitutes a central value of society capable of economic, cultural and spiritual interpretation. Translated into the design of human settlements, efficient energy use produces small, dispersed, semi-autonomous centres of perhaps 20,000 to 30,000 people arranged in clusters – village and small-town stars, as it were, in a city constellation. High densities and centralization are inefficient. So is land-use 'apartheid', the rigid separation of home, work and pleasure through planning regulations.

The pattern described here bears a striking resemblance to pre-industrial and post-industrial models. Peter Laslett, for example, describes seventeenth-century England as a network of independent communities linked not only by personal connections but through local rural centres of exchange and communication. It was, he says, a 'reticulation rather than … a particulation – a web spread over the whole geography'. Further similarities are recognizable both in Ebenezer Howard's social city and in the picture of a post-industrial society fragmented into a vast lacework of self-supporting rural hamlets predicted in works like *Europe 2000*.

It is also, of course, not merely the pattern of Milton Keynes or of Lightmoor but one remarkably close to that emerging in the old rural areas most affected by the outflow from the cities: the 'necklace' of electronic villages around Cambridge, for example. Yet the achievement of energy-efficient settlements is vitally dependent on smoothly functioning private land markets. In Milton Keynes the achievement was relatively easy: the land was public. Around Cambridge conflicts over green belt, farmland and structure plans have led to rigidities and blockages.

The implication of this argument is that the alignment of energy-conserving settlement patterns with apparently unconnected human or social goals is not coincidental: that it represents, through economic or cultural pressures, the steady diffusion of the philosophy of ecological design. Energy is an ecological imperative, the fossil fuel crisis a kind of planetary distress signal, and in responding to it human beings' image of themselves and their role in

Plate 13.1   Milton Keynes: the woodland belts of the parkway system

the universe has radically altered. One consequence is that they wish to live and work in places which allow them to express ecological awareness. There is, in other words, not merely an emotional search for 'harmony with nature' but a more overtly intellectual search for an appropriate social setting.

Case law makes the point better, however, and Peter Laslett has provided this in his vivid depiction of the pre-industrial link between marriage age, population growth, social stability and what Laslett calls a 'slot' but might equally well be called an ecological niche. Such slots or niches consisted, say, of a cottage with land attached, or a small business. Couples had to wait for them to fall empty before they married and produced children. If there were no vacant slots, queues lengthened and fertility fell.

Here, says Laslett, was a society 'responding rhythmically' to its environment. Here, too, for ordinary people, was a powerful dramatization of the concept of environmental carrying capacity – plainly visible to them, indeed inescapable, by virtue of the sheer miniaturization of sixteenth- and seventeenth-century society. In a mass society such perceptions and connections blur into invisibility.

Ecological design – design with nature – is a fundamental principle of land management, and by extension management of the earth, its landscape and resources. It seeks to rejoin functions and identities disjoined by industrialism. It treats human beings as a vital element in the ecological mosaic. As ecodevelopment, or sustainable growth, it means advances in human welfare which are kind to the environment. As creative ecology, it means a shift in conservation from defensive heroics to the more subtle and constructive skills involved in the design and management of ever larger tracts of landscape. In practical terms it means evaluating the impact of human development and then 'building in' the green dimension. It thus means more spacious, ecologically rewarding, productive and energy-efficient settlements. These are, of course, exacting specifications. Fortunately, they are being translated into practice with increasing, although many would argue still inadequate, urgency and skill.

The greening of the cities was a watershed, the moment when design with nature was first applied across a broad, popular front to a way of life that most rigidly embodied the old exclusions and

Plate 13.2　Country lanes in a new housing landscape at Warrington-Runcorn new town

divisions. The essence was small-scale experimentation by individuals and groups. But the philosophy of ecodevelopment increasingly influences larger operations.

The Rother Valley country park, opened in 1983 on 750 acres of former tips and jaded scrubland in built-up South Yorkshire, is one such venture: a ten-year project combining coal extraction, flood relief, recreation, woodlands and wildlife conservation. The final landform was agreed *before* the National Coal Board began work. The River Rother was diverted and four lakes created, doubling as washlands. Beds of reed and reedmace prevent bank erosion and sieve out many polluting solids when the Rother is in flood. Perhaps the most impressive feature about Rother Valley is not that flood relief and coal extraction made a country park possible – as a pure recreation proposal it was considered prohibitively expensive – but that it imposed design and coherence on what had previously always happened through careless development, ignorance and neglect. Those were the forces that had produced the unofficial countryside of gravel pit, sewage farm and urban wasteland. At Rother Valley, in effect, men substituted order for accident and took a decisive new grip on their destiny. In 1984, incidentally, the British conservation lobby set up the first national organization specifically to promote ecodevelopment.

Much has also been learnt about the design of cities with nature. From the energy-conserving house of Milton Keynes it is a lesser step to Bristol's future city home and from there to the bioshelter, a small productive ecosystem of people, plants and fish ponds in which living room and garden merge into one.

The bioshelter is the ultimate vernacular: a small chunk of the natural landscape adjusted for human living. Its technological sophistication is dedicated to the single end of solar thrift. In the case of biotecture or 'vertical gardening', which has transformed many city-centre façades and rooftops in Zurich, Dusseldorf and Stuttgart, the building actually 'breathes with the leaf'. A carpeting trelliswork of evergreen, the leaves raising themselves in summer and lowering themselves in winter, has cut energy bills by up to 30 per cent.

Bioshelters and their counterparts – examples include Sweden's Naturhuset, the 'integral urban house' of California's Farallones Institute, West Berlin's Oekotop project – are still primarily experimental. But they represent the most promising avenue of innovation in house design. In this sense the Oekotop venture is the most challenging, since it involves a city block of 500 apartments and

1,200 people where technologies have included rainwater collection, food production on roofs, on-site sewage treatment and comprehensive waste and water recycling. At Oekotop environmentalism has demonstrated its fundamental role in community development.

At the level of strategic planning we also know that, for the greatest richness and diversity of wildlife, elongated stretches of greenspace are preferable to circular areas, acting, for example, as interceptors of birds and insects, and that one large green area offers more opportunities than several smaller ones. With birds, for instance, the relationship between area of greenspace and species variety is logarithmic. Green corridors linking larger rural spaces reduce the effects of isolation. Much of the acquired wisdom, most notably on the importance of 'fauna-roads' for migrating animals, has been brought to bear on an unusual experiment in Poland. Transport routes cause immense disruption to wildlife movement: ring roads and bypasses turn towns into fortresses and produce corpses by the thousand. At Bialoleka Dworska, a new suburb of Warsaw, underground passages below main roads allow mammals easy circulation between settlement and adjoining forest, free from the threat and noise of traffic.

Yet the most persuasive reason for the design of cities with nature remains human need. Landscape preference studies now offer firm evidence that the greening of cities brings great psychological benefits: that it enriches lives and restores lost connections. One American study found such a numbed hostility to the built environment amongst city-dwellers that it concluded, 'the natural setting seems to offer ... a quality of experience for which the city provides no substitute.' Even more striking was the 'sensory mapping' exercise in Ashland, Wisconsin, on the shores of Lake Superior, where residents were asked which parts of the town best expressed its identity and their own. What made them feel most 'at home'?

Ashland residents expressed themselves overwhelmingly in favour of natural landscape features. These provided 79 per cent of all valued sites: the most disliked areas were 'constructed-urban'. Eighty-three per cent of the features that best summed up Ashland's sense of place and 69 per cent of those that made them feel at home were natural. And 76 per cent of the most memorable sensory experiences were associated with 'primitive-natural' landscapes. Some of these experiences – the sounds of wildlife, the smell of fresh air – were perhaps predictable. But the researchers noted the less obvious associations. 'Cold weather, snow, rain, lake winds, the changing visual qualities of the lake, wildlife and change of seasons,' they observed, 'were some of the elements of primitive ... nature most highly valued by Ashland city residents.' The city, in short, was loved for what it was not.

Design with nature provides the prospectus for a new programme of resettling the countryside. New constellated cities or village clusters could create a grazed, coppiced, walked-through, lived-in and altogether better husbanded countryside richer in wildlife than the empty farmland or drab green belt they replace. Reuniting jobs and people in more spacious settlements would constitute a measure of enormous social efficiency, saving energy, enhancing local self-sufficiency and pride, and opening up new prospects of home-centred economic revival. Places, particularly if they were co-operatively designed and built and thus achieved a balance between private and communal space, would be better-loved, their laws and institutions commanding readier assent. This is the programme that has emerged in the free space of cities and now struggles desperately for a foothold in the countryside.

The strength of this programme has been recognized in some unlikely quarters – most notably in the proposal by Consortium Developments, representing the volume house-builders, for up to fifteen new villages around London, each of between 13,000 and 18,000 people. The plans for the first of those villages, at Tillingham Hall on the dull agricultural flatlands of Essex, envisaged lakes, nature reserves, green corridors and extensive tree-planting and landscaping. In environmental terms, a rich landscape would replace a poor one. The proposal was, predictably, rejected in 1987.

An obsessive focus on the quantity of countryside lost to 'urban sprawl' has clearly obscured issues of quality. There is evidently a difference worth noting between ugly settlements on beautiful countryside and beautiful settlements on ugly countryside. But the land-use equation has to be adjusted even further if we are to confront the issues of productiveness and personal space.

Britain's impending farmland surplus may not prove immutable. Indeed, the longer-term realities of the international food order, impinging when EEC-inspired overproduction is little more than a memory, indicate that self-sufficiency, whether local or national, will increase rather than diminish in importance. The historical trend towards higher yields from land, retarded by capital-intensive agriculture, will thus need to be enhanced. Conventional farming clearly cannot manage this without unacceptable costs in terms of environment, energy and human health. But even if it could, the evidence points compellingly to the need for an alternative.

In practice this means enlisting an army of small farmers, market gardeners and horticulturalists, prepared, probably keen, to grow promising 'alternative' crops – trees like the black walnut, for example, are up to 400 times more productive than wheat – and to experiment with a mixture of traditional polycultures and new permacultures. And this, in turn, may mean widening the concept of the housing estate into the smallholding demesne, an echo of the pre-industrial manorial system.

Vital questions of space and density are involved. Bill Mollison estimates that for food production a quarter of an acre is ideal, a house-plot size common in Australian cities but not in Britain, where more than half of all houses with gardens stand on less than one-sixteenth of an acre. During the 'great gardens controversy' of the 1950s it was found that the typical housing estate, with 14 per cent of its area cultivated, produced as much food as the 'better than average' farmland it replaced. More crucially, it emerged that bigger gardens encouraged householders to grow their own food. The sharpest production increase came at about a tenth of an acre.

The implications of this are worth pursuing. Let us notionally allow every household in the country one-tenth of an acre – a plot roughly 100 feet long and 40 feet wide – and assume a standard urban relationship between net residential area and other land-uses. Both, incidentally, are 'worst case' assumptions. Such apparently lavish provision, far from enormously enlarging the urban area, would cut it by 20 per cent, almost a million acres. The food produced, moreover, would be fresh and relatively free of chemical residues: few people consciously poison themselves. For a small investment in land, a rich new source of employment would be generated, although not in the form of conventional wage-paying jobs.

The calculation illustrates both the relationship between land-use and new work practices and the inequities of current patterns of land ownership. It also demonstrates the baselessness of so many worries about land loss to the demon of urban sprawl. Britain devotes more of its land to farming than any other developed country yet gets much less food from it; the culprits are large farms and extensive mechanized farming. Its urban densities are extremely high, second only to Italy in the developed west. Contrary, again, to widespread misapprehension, vast quantities of farmland have not disappeared. The urban area, into which virtually the entire population is crammed, has grown from 8 to 11 per cent since 1939 and will rise to 12 per cent by the year 2000 if housing needs are met.

These gross strategic inequities, which for so many city-dwellers translate into high-rise misery, are underpinned by a system of green belts, development control and land-use apartheid which raises land prices, produces smaller rooms in smaller houses and makes home ownership, for millions, a much-desired impossibility. It also denies them those most common, innocent, necessary and thus most comprehensively ignored of human activities, walking and gardening, occupations in which profit, pleasure and creativity make for better-adjusted individuals and thus a better-adjusted society.

Green belts, for example, are poor in terms of public access, little visited by inner-city residents, provide few recreation sites and are no substitute for the countryside. Over a third of green belt land is derelict, shabby or agriculturally run-down. Green belts do not halt development: they merely turn it into a piecemeal, conflict-ridden and time-consuming outwards creep of the city. From the point of view of preserving the identity of expanding settlements, the original aim of the green belt, the Greek solution of throwing off a new colony remains immeasurably superior.

This, of course, was what the garden city pioneers sought and what the first New Towns programme, conceived half a century ago, sought to achieve: the dispersal of city populations. In the event it was overtaken by massive post-war population growth. The New Towns, in effect, served as an overflow channel for the demographic flood. The cities remained congested. Relative population stability thus allows us an unprecedented historical opportunity: that of rationally relating our population size both to individual choice and to the environment's carrying capacity.

Some ecologists have already attempted this for *homo sapiens*. Eugene Odum, for example, calculates that 50 per cent of man's living environment should be natural. Beyond this 'critical mass', he argues, diseconomies set in. Many potentially successful settlements have indeed been destroyed by sheer weight of numbers. The medieval European town was one example. Another was the Bronx in New York, where Olmsted's early greenways and open-space planning were simply swamped by population influx.

Since the strategic objective of the first great programme of urban dispersal remains largely unfulfilled, however, common sense indicates that we should try again. The reasons are compelling:

employment patterns, community development priorities, social equity and efficiency and, above all, individual preference – for habitable cities and inhabited countryside.

A second 'New Towns' programme will be very different from the first. That much is evident from the extended preview of it which has been playing in the cities to large audiences. The urban greeners are the potential colonists of the countryside: the electronic peasants and ecocrofters and wild gardeners of some future agropolis. The city of the 1970s and 1980s was thus a dress rehearsal for the post-industrial resettlement of the land, a programme that goes beyond New Towns to new villages, new farms and a new nature.

There is, however, a wider significance. In the concept of agropolis – the revitalized and productive rural settlement through which China sought to resolve one of the three great contradictions diagnosed by Mao Tse-tung, that between city and country – lies the developmental convergence of global North and South, the place where the tensions and pressures of aid and trade vanish. The Chinese themselves have been assiduous both in greening the cities and in revivifying the countryside. But the resolution of Mao's contradiction, embodied in the city-that-is-countryside or the countryside-that-is-city, makes both terms redundant. As with the emergence over a century and a half of the planetary city, from metropolis to megalopolis to ecumenopolis, the logic is simple. If the city is everywhere, it is also nowhere.

Hence the vision central to the greening of the cities, of a city that has vanished and a countryside that is lived in and lived off, may mark the culmination of that centuries-slow overwhelming of *terra incognita* by *oecumene*, the encircling of wilderness by settlement, and its transition to a new stage. In that stage man 'takes in the earth at one view', a perspective reserved throughout most of history for God alone, and devolves upon himself the task of planetary management. In the cities he has learnt to reclaim land, reintroduce farming, and tempt back fellow creatures. But increasingly he experiments on a larger scale.

In Scandinavia, outside any park or wildlife reserve, the wolf has returned, to conservationists' delight and farmers' alarm: one specimen savaged 140 sheep. In India the Bengal tiger has been assigned 10,000 square miles in fifteen reserves – but forty villages were removed and 400 people have been killed. In California the struggle to save the condor seems doomed. In Zimbabwe, where people clamour for new land and crowd wild animals into ever smaller game refuges, there is more cause for optimism. The land-use plan for Sebungwe, 15,000 square miles of subsistence farmland surrounding prime wilderness areas, envisages a shift away from 'pure' conservation. Instead the better soils will become farmland, the poorer soils wildlife areas. In between will be buffer zones on which controlled exploitation of wildlife will be allowed, the profits accruing to local people. Longer-term thinking, however, here and in Sweden, centres on the need for 'core' wildlife refuges connected by corridors for animal migration: a sub-continental design that parallels and amplifies the city's emphasis on greenways and blueways, on wildlife reservoirs, corridors and stepping-stones.

These solutions are imperfect because the divine role is a hard one to fill, but there is no alternative. We cannot afford to apply to the planetary city those same doctrines of human exclusiveness that desolated its precursors and ravaged their hinterland. The greening of the cities, and the linked resettlement of the countryside, is thus a long-term disaster-avoidance strategy. More than that, however, it presages a decisive transformation of human perceptions, for since the boundaries of the city are now the boundaries of the planet, 'wilderness', the *terra incognita* once the abode of monsters and miracles, no longer lies outside and beyond human settlements. To seek or promote it there, where it has vanished, is not merely to institutionalize psychological tension and thus to pre-ordain social conflict but, which is far worse, to build a rhythm of frustration and neurosis into the deepest structures and patterns of human society: to waste time and spirit looking for what we know we cannot find. And to persist in that search, so that it becomes merely wilful and self-indulgent escapism, is to court a pornography of the natural world, voyeuristic fantasies of the wild and the primitive which are likely in the long run to prove even more corrosive of human values.

The truth is that city and planet do not merely have an objective, geographical existence: they are not merely 'things-in-themselves'. They are also image-objects of inner space, and around them, a solitary traveller in a vast mental universe, moves the small focused point of human consciousness. And since what happens in the outer world of settlements and wilderness does not merely describe or represent, but in a vital sense actually constitutes what happens in the inner world of the mind, the thwarted quest for wilderness marks a crossroads in the evolution of consciousness, a turning point comparable to the discovery of tools or the invention of abstract thought through language. The horizons of the mind have been located. Beyond them is not wilderness but inter-stellar emptiness, an altogether different quality. We must retrace our steps or fall off the edge. Like the Sioux Indians or the Bushmen, we must make all nature our home.

Hence it is of immense significance that in the greening of the cities people have been able to rediscover the miraculous and the marvellous among the common, the ordinary and the near-at-hand. If a garden is a game reserve and a city park a forest, distant latitudes lose their allure. If the green that surrounds us is reanimated, grows instinct with the numinous, we live comfortably with our mysteries. In both cases, whether through a deep-rooted instinctual reflex or a clever piece of self-deception, we appear to have contrived or connived at this revitalization of the home ground. Yet by aligning our skills with what we portray to ourselves as deeper rhythms and realities, we have done so in a way that makes illusion inseparable from reality: the trick of landscape art since its beginning – the trick, also, of religion.

Jung, it seems, was wrong. We *have* made our own gods and they surround us daily. In so doing we have demonstrated great qualities of maturity and flexibility and great reserves of inventiveness. The urge to escape remains. But in showing that we can reconcile and transcend it, that we can 'escape inwards', bring wilderness back into our homes and our minds and our settlements and thus make it something more than wilderness, we have planted a small seed of hope for the future.

NOTES

To avoid duplication, sources of quotations etc. are only specified in the notes where they cannot be immediately identified from the bibliography, for example, where an author or organization cited in the text has two or more bibliography entries. In this case the notes specify the relevant bibliography entry, with the date of publication in parentheses. Where there is no reference in notes or bibliography, the source is interview, personal communication or ephemera such as information leaflets.

CHAPTER 1 • DETRITUS

'Absolute' job loss in London: Chisholm; Weatheritt and Lovett; Foster and Richardson. The last two say respectively, 1961 and 1962.

Early London: Rasmussen.

Decentralization: Hall (1973).

Job losses: Hausner; Thomas, D. (1983b); GLC (1981, 1985); Martin, R.L.; Martin and Hodge.

Company births and deaths: Cameron, Chapter 12; Elias and Keogh.

New towns and inner-city job losses: Fothergill *et al.* (1982–3, 1983).

Office decentralization: Jones, Lang, Wootton.

Economic upturn would accelerate decentralization: see Fothergill *et al.*, op. cit.

Link between size of place and pace of decentralization: see Census; Chow; Champion; Keeble (1980); Hall (1973), p 250.

Remoteness: Champion; Long.

Counterurbanization: Berry (1976); Morrison and Wheeler; Vining and Kontuly; Hall and Hay; Fielding; Champion; Hodge, G.

Counterurbanization in Europe: Fielding.

Urban-rural manufacturing shift: Keeble (1983).

Cornish survey: Town and Country Planning, October 1983.

Footloose industry: see Oakey for study of 102 high-tech plants which support this argument.

Constrained location theory: Fothergill *et al.*

Suburbanization: For evidence that different forces are now at work, see, for example: Long; Champion; Morrison and Wheeler; also Vining, D.R. and Strauss, A., 'A demonstration that the current deconcentration of population within the United States is a clean break with the past', *Environment and Planning*, A,9,7, 1977.

Collapsing infrastructure: Pearce, F. and Hamer, M., 'The empire's last stand', *New Scientist*, 12 May 1983; Institution of Civil Engineers.

Housing repairs: £47 billion estimate by Association of Metropolitan Authorities, quoted in *The Times*, 21 October 1985.

Alice Coleman's figures: see, for instance, Coleman (1980).

Land figures: DoE (1984); RSA (1980, 1981); Burrows (1978); West Midlands CC, *Derelict land in the West Midlands*, June 1983.

Vacant commercial property statistics: King & Co.; Jones, Lang and Wootton; DoE/Welsh Office, Commercial and industrial floorspace statistics.

Commercial vacancy: Industrial statistics from King & Co, office statistics from Hillier, Parker, May and Rowden, quoted in the *Financial Times* Special Report on office property, 18 May 1984. Buildings of less than 5,000 and 20,000 square feet, respectively, are excluded. King & Co. figures also exclude semi-derelict premises. Centrepoint is 200,000 square feet.

Empty houses: Figures, from DoE and Empty Property Unit, based on Housing Investment Programme returns and DoE census estimates. Land equivalents based on Nationwide Building Society average site area data (3,399 square feet for conurbations, 4,295 square feet nationally).

Land market: RTPI. For evidence that gentrification is likely to prove highly limited, see Hamnett; Spain, D., 'Black-to-white succession in central city housing: limited evidence of urban revitalization', *Urban Affairs Quarterly*, 15, 1980.

Arunta: Quoted in Lynch (1960).

Medieval guilds: Braudel.

Tudor rural renaissance: Everitt.

Housebuilders' survey: Nicholls *et al.*

Hall survey: Hall *et al.* (1973).

Size and dissatisfaction: Dahmann; US Bureau of the Census; International Labour Office. For the increase of crime with size, see also Fisher, C.S., 'The effects of urban life on traditional values', *Social Forces*, 53, 1975.

MORI poll: CDP.

Countryside selling itself: Burgess.

CHAPTER 2 • THE GREAT WEN VERSUS THE GARDEN CITY

London's growth: Weinreb and Hibbert; Olsen; Trent; Hall *et al.* (1973); Braudel.

Cities, gods and kings: see Mumford; Kramer.

Kinglist: laws and traditions governing kingship, as set down in Sumerian poetry: see Mumford; Kramer.

London's 'distancing' of other towns: Hoskins; Fisher; Dyos.

Christaller and Jefferson: Carter.

Grid-plan cities: Stanislawski, S., 'The origin and spread of the grid-pattern town', *Geographical Review*, 36, January, 1946; Hawley.

Land under feudalism and capitalism: Harvey; Ryan; Becker; Gillie.

Haig and 'frictionless' cities: see Lampard, E.E., 'The history of cities in the economically advanced areas', pp.81–137 in Redfield *et al.*

Economics and ecology: ecology, according to H.G. Wells and Julian Huxley, was 'biological economics … an extension of economics to the whole of life'. The Chipko (Indian tree-hugging) slogan is 'Ecology is permanent economy.'

Norwich: Hoskins.

German cities: Braudel.

Rome, soil erosion and rural stress: Hughes.

Weber: quoted in Mumford.

Green belt growth: Munton.

Britain and the Beast: Williams-Ellis.

London as metropolis: Dyos.

Super-metropolitan areas: Hall (1984).

Super-megalopolitan regions: Hall and Hay.

Rapoport on Boswash: in 'Environmental quality, metropolitan areas and traditional settlements', *Habitat International*, 7, 3/4, 1983.

Hall on megalopolis England: Hall *et al.*

Redefining city: SSRC (now ESRC) project was undertaken by the Centre for Urban and Regional Studies at Newcastle University. Similar research was started in 1980 by the OPCS and the DoE. The assumptions made in these studies are fascinating, especially in relation to the concepts of boundaries and enclaves (see below, Chapters 11–13). See OPCS, 'Key statistics for urban areas', Census, 1981; HMSO, 1984; 'Population trends', *Urban Britain*, Summer 1984, pp. 10–18; Champion, A.G., *et al.*, 'A new definition of cities', *Town and Country Planning*, November 1983. On conservatism in urban planning, see Cherry (1972).

CHAPTER 3 ● WILDERNESS, NATURE AND MUNICIPALITY

Change in attitudes to city: Williams (1983).

Eliade: (1982).

Cosmos and living body: the most influential statement is Plato's in the *Timaeus*.

Buffon: quoted in Glacken.

Johnson and the Highlands: Williams (1975).

Loudon and parks: *Landscape Design*, 140, 1982.

Select committee of 1840: Laurie, M., 'Nature and city planning in the nineteenth century', in Laurie, I.C. (1979).

Parks' role in reducing discontent: Walker and Duffield; Chadwick.

Victoria Park: Rendel, S., 'Renewal of parks in inner London', *Landscape Research*, 8, 2, Summer 1983.

118,000 in Victoria Park: Walker and Duffield.

Pope and Walpole: Hunt and Willis.

Gustavsson: quoted in Ruff and Tregay.

Gans: quoted in Ruff.

'Value-laden landscapes': Ruff and Tregay.

Shaftesbury: in *The Moralists*.

Pope: Hunt and Willis.

Boston and the City Beautiful: Laurie, M., op. cit.

American wilderness movement: Nash.

Thoreau: quotation from 1851 lecture (see Nash). On his proposed wilderness areas, see Laurie, M., op. cit.

Olmsted: quotation from *A Consideration of the Justifying Value of a Public Park*, Boston, 1881.

Colorado river trip: Nash.

Thoreau: quotation from his essay, 'Walking'.

Hopkins: quotation from 'Inversnaid'.

CHAPTER 4 ● THE RECOVERY OF THE PRIMITIVE – ENERGY, ECOLOGY AND GOD

Environmentalists, politics and the church: the estimate of three and a half million is a revision of Lowe and Goyder's figure of three million made necessary by growth in environmental groups. Four organizations alone – the National Trust, the Royal Society for the Protection of Birds, Greenpeace and the Royal Society for Nature Conservation – have added at least 485,000 to the Lowe and Goyder figure (*Social Trends*, London, HMSO, 1985). The combined individual membership of the Conservative, Labour, Liberal and Social Democratic parties is 1.68 million (*The Times*, 9,16,30 September, 7 October 1985). Average Anglican Sunday church attendance in 1983 was 1.12 million (*Church Statistics*, Church House, London, 1985), down from 1.24 million in 1980. Official church statistics include children and may overstate actual attendance by a factor of two and a half (Francis, L., *Rural Anglicanism*, Collins, 1985).

Eliade: (1968).

Autochthony: ibid.

Fusion of separate traditions: see the 'new alliance' between science and humanities spoken of in Prigogine and Stengers. Also Sheldrake; Hardy (1975).

Descartes: quoted in Sommers, F., (1978), 'Dualism in Descartes: the logical ground', in Hooker, M., (ed.), *Descartes*, Johns Hopkins University Press.

George Sessions: 'shallow and deep ecology: a review of the philosophical literature', in Schultz and Hughes.

Tolkien: the association of *The Lord of the Rings* with the cultural forces behind environmentalism is suggested in the way the book suddenly caught on, particularly amongst the student generation, in the mid-1960s, after having long remained the property of a small coterie. In 1966 it headed the list of paperback best sellers.

Animal cries: quotations from Eliade (1968), p. 61.

Animism among Bushmen and Aborigines: Martin, V.

Cave paintings: quotations from Levy. See also Maringer.

Robins, hares: Thomas, K.

Solar output, biomass figures: Owen, D.F., p. 113.

Logos spermatikos: Glacken.

Taoism: Needham.

Geomantics: quotation from Needham. See also Pennick.

Delaware study: Mudrak. Also Ulrich, R.S., 'Visual landscapes and psychological well-being', *Landscape Research*, 4,1,1979.

Post-operative patients and trees: Sunday Times, 24 June 1984.

Geochemistry: paper given by Jane Clark at the British Association for the Advancement of Science conference, 28 August 1985.

Sunshine therapy: New Scientist, 26 July 1985.

Negative ions: Soyka; *The Times*, 4 September 1980; Jones, D.P., *et al.*, 'Effect of long-term ionized air treatment on patients with bronchial asthma', *Thorax*, 31, 4, August 1976; Krueger, A.P. and Reed, E.J., 'Biological impact of small air ions', *Science*, 193, 24 September 1976; Hawkins, L.H. and Barker, T., 'Air ions and human performance', *Ergonomics*, 21, 4, 1978.

Hippocrates and the humours: Glacken.

Mana and its analogues: Eliade (1968).

Lindeman: see also Owen, D.F., p. 122.

Diversity, richness and plenitude: see, for example, Hutchinson, G.E. (1959), 'Homage to Santa Rosalia, or Why are there so many kinds of animals', in Kormondy.

Gabon botanic list: Lévi-Strauss, p. 5.

Pawnee Indian: ibid., p. 10.

Circuit diagrams: Odum, E.P. See also Odum, H.T. (1967, 1971) for elaboration of these ideas.

Leopold: quotation from essay 'The land ethic'.

McHarg: cf. Hardy, A.

Eliade: (1968).

Gaia hypothesis: Lovelock, J. and Epton, S., 'The quest for Gaia', *New Scientist*, 6 February 1975.

The prophet Smohalla: quoted in Eliade (1968).

Rights for trees etc.: Stone, C.D., *Should Trees Have Standing? Towards Legal Rights for Natural Objects*, William Haufman, 1974.

Witchcraft: Tysoe, M., 'The great British witch boom', *New Society*, 18 October 1984. See also Adler.

Teilhard's 'mind-layer' and evolutionism: Towers.

CHAPTER 5 • A GEOGRAPHY OF THE SACRED

Jung on UFOs: (1959b).

Münster's Cosmography: Glacken.

Marvels of the East: McGurk, P. *et al.*, *Early English Manuscripts in Facsimile*, vol. 21, Introduction, Allen & Unwin, 1984.

Cynocephalus: Ibid.

Cartography: Bagrow; Crone; Tooley.

Aquinas on Paradise: Glacken.

Columbus and Haiti: Eliade (1968).

Tillyard: p. 45.

Maps and reality: Robinson and Petchenik; Oatley; Keates.

Aleutian islanders: cited by Lynch (1960) from Elliott, H.W., *Our Arctic Province*, New York, 1886.

Pope: Tillyard, p. 34.

Forest cover: Best, pp. 10–11. The current figure is 9 per cent: the increase is due largely to conifer-planting.

Gardens and enclosed space: Batey, M., 'The evolution of the English landscape park'; Watson, J.R., 'Parkland in literature'. Both in *Landscape Research*, vol. 7, no. 1, Spring 1982.

Japanese garden: Watts.

Outdoor rooms: Hunter.

Planet earth: for a provocative discussion of how this image has affected social and cultural attitudes, see Russell, P.

Continental drift: The Times, 5 September 1984.

American frontier: Turner quoted from the *Atlantic Monthly* in 1896, cited by Nash.

Leopold: p. 158.

God and Einstein: Davies.

Jung: quotations, respectively, from Jung (1953); Fordham, p. 27; and Jung (1959a) – latter in essay, 'Archetypes of the collective unconscious'. See also Mitchell, J.G., 'Why we need our monsters', *National Wildlife*, 16, 1978.

'Going, Going': in *High Windows*, Faber & Faber, 1974.

Browne: in *Religio Medici*.

Jung: quotation from (1959b).

CHAPTER 6 ● THE PARABLE OF THE BOG

Because of the sheer number of sources of information in this chapter and those that follow, references are included only where a point seems to need reinforcement or a quotation attribution.

Tolkien's memories: Carpenter (1977).

Kant and noumena: Russell, B., pp. 680, 685; Hart, R.A., and Moore, G.T., 'The development of spatial cognition: a review', in Downs and Stea.

Landscape and the classical poets: see, for example, Hunt and Willis.

Farmer as landscape architect: Lowenthal and Prince.

Abercrombie and urbanism: in Abercrombie, P., *Town and Country Planning*, Butterworth, 1933.

Communing with nature: Shoard, M., 'The lure of the moors', in Gold and Burgess.

Wilderness-users' survey: Hammett.

Children, the elsewhere schema and domicentricity: Lee, T.R., 'Psychology and living space'; Hart and Moore, op. cit.; both in Downs and Stea.

Housing and land prices: see, for example, *Homes, Jobs, Land – The Eternal Triangle*, House-Builders' Federation, 1985; Hall *et al.*; Commons Select Committee on the Environment vol. 3, p. 564. Calculation based on Hardy and Ward; CSO; London and Cambridge Economic Service; Burnett; Butler and Sloman.

Smaller houses, gardens: Hall *et al.*

'Trapped' city-dwellers: DoE (1977); Elias and Keogh.

Countryside Commission Survey: Shoard (1979).

Dower: quoted in Shoard (1979).

Buber: cited by Eliade (1968).

SSIs, ecology and objectivity: see Adams, W.M. and Lowe, P.D., 'Continuity and change: science and values in nature conservation strategies', *Discussion Paper* no. 36, British Association of Nature Conservationists/ University College, London, 1981.

CHAPTER 7 • THE COUNTRY COMES TO TOWN

Wildlife reports: the best sources of these are the publications of the various urban naturalists' groups.

Wasteland flora: see, for example, Davis.

Urban commons: Gilbert, O., 'The wildlife of Britain's wasteland', *New Scientist*, 24 March 1983.

Green space, gardens and cities: figures from Davis; Best, p. 68; Sukopp and Werner, p. 17.

King: cited by Best.

Thoreau and Mt Katahdin: *The Maine Woods* (1864).

The prairie: Evernden, N., 'Getting across the prairie', *Landscape*, 27, 3, 1983. For analysis of a comparable British landscape, see Burgess, J., 'Filming the fens: a visual interpretation of regional character', in Gold and Burgess.

Elton: quoted in Bornkamm.

City commons: Gilbert, op. cit.

Regressing landscapes: see, for example, GLC (1984).

Highgate cemetery: Jenny Cox's landscape and vegetation report was commissioned by the Friends of Highgate Cemetery and published in 1976.

Cities as 'natural' habitats: see, for example, Sukopp and Werner; also Nicholson, M., 'Life in the city', *Naturopa*, 36, Council of Europe, 1980.

Urban ecosystem: Douglas; Landsberg; Oke.

Birds 'tamer' in towns: Cooke, A.S., *Biological Conservation*, 18, 2, 1980, pp. 85–8.

Pygmies, farmers and forests: Martin, V.

Sioux and nature: quoted in McLuhan, T.C. (ed.), *Touch the Earth: A Self-Portrait of Indian Existence*, Outerbridge, 1971.

Plant gains and losses: Davis.

Farms, housing and wildlife: Davis; Ratcliffe. See also Chapter 13.

British Rail: Sargent, C., *Britain's Railway Vegetation*, Institute of Terrestrial Ecology, Cambridge, 1984.

Urban fringe: Countryside Commission (1976).

Urban conservation groups: see 'Enterprise profile', UK Centre for Economic and Environmental Development, *Bulletin*, 6, November-December 1985.

Chris Rose: *Countryside Campaigner*, July/August 1983.

Lovell Construction: quoted in *Weekender* (local paper), 14 May 1982.

Experiences of nature in cities: Mostyn.

W.H. Hudson: *Birds in London* (1898).

CHAPTER 8 • A PEOPLE'S LANDSCAPE

Rural Preservation Association: see, for example, *Greensight: Four Years of Native Planting and Habitat Creation in Liverpool*, RPA, 1984.

Nature in cities symposium: see Laurie, I. (1975, 1979).

Netherlands: Laurie, I. (1979); Ruff; Tregay (1980); Westpal, J., 'Almere, New town', *Landscape Architecture*, 72, 3, May 1982; Bos, H.J., 'The ecological design of urban green space in Holland', *Landscape Research*, 6, 3, Winter 1981.

Berry: in (1973).

Le Roi: quoted in Ruff (1979).

H.J. Bos: in Bos, H.J. and Mol, J.L., 'The Dutch example: native planting in Holland', in Laurie, I. (1979).

Institute of Preventive Medicine study: Tregay (1980).

Andries Vierlingh: quoted in Glacken.

Piaget: Hart and Moore; Piaget and Inhelder (1967, 1969); Ward.

Lower Swansea Valley: see, for example, Lavender.

Trees and insects: Southwood, T.R.E., 'The number of species of insect associated with various trees', *Journal of Animal Ecology*, 30, pp. 1–8, 1961; Moreton, B.D., *Beneficial Insects and Mites*, Ministry of Agriculture, Food and Fisheries, HMSO, 1969.

Pocket parks, etc.: The new 'home-made' landscapes of the cities are chiefly described in the literature of environmental voluntarism, which with a few exceptions – the work of the National Council for Voluntary Organizations, for example, or the campaigns sponsored by Shell, Ford and Kodak – is simply too voluminous and fragmented to summarize.

William Curtis: see Hadfield, p. 253.

A 'new type of open space': quotation from the Ecological Parks Trust's second report, in September 1982. See also Cotton, J., 'Field teaching in inner cities: the William Curtis ecological park', *Journal of Biological Education*, 13, 4, 1979.

Free Form: a full account of the making of the Daubeney Road community arts garden is in the *Bulletin of Environmental Education*, 125, October 1981.

CHAPTER 9 • SMALL WORLDS

Insect recordings: see also Owen, J., 'The most neglected wildlife habitat of all', *New Scientist*, 6 January 1983; and Davis.

'Wild' garden in history: see, for example, Hadfield, pp. 168, 362; and Hunt and Willis.

George Eliot: Middlemarch, Chapter 22.

Minibeasts: see, for example, Stoker, B., 'Areas for minibeasts: Action Pack', Cheshire Education Authority, 1980; *School Garden Book* (translated from the Dutch *Schooltuinboek*), Rural Preservation Association, 1984.

School and society: see CUSC; Ward.

Field studies: CUSC.

Washington school yard: Moore, R., 'Post-hoc prospectus on the environmental yard', *Bulletin of Environmental Education* (BEE), 124, August-September 1981; Moore, R., 'Anarchy zone', *Landscape Architecture Quarterly*, 65, 5, 1974; Hart, R., 'Wildlands for children', reprinted from *Landschaftstadt*, in BEE, 141, February 1983.

Nature as outdoor classroom: Ward; Schmitt. See also Sames; 'Schoolyard landscape' (papers presented at a seminar of the Yorkshire and

Humberside chapter of the Landscape Institute, 14 April 1981); Fry, G., and Wratten, S., 'Insect-plant relationships in ecological teaching', *Journal of Biological Education*, 13, 4, 1979.

South Yorkshire schools: BTCV, South Yorkshire schools project, 'Report', 1982–83; Canning, M.A., 'School nature reserves in South Yorkshire', *Landscape Research*, Spring, 1980; South Yorkshire CC, 'Statistics of reclamation, environmental improvement and conservation schemes, 1974–84'.

Apathy and cities: quoted from ERC's annual report, 1980–81.

The work of Cornell and the ERC: described, for example, in White, G., 'Learning to change', *BEE*, 125, October 1981.

Acclimatization: the basic source books are van Matre's *Acclimatization* and *Acclimatizing*, Acclimatization Experiences Institute, Warrenville, Illinois/ YMCA National Centre, Ulverston, Cumbria. AEI is a non-profit-making educational organization founded in 1974.

Peas and pods: *Guardian*, 16 February 1981.

Milk, etc. from animals: *Farmers' Weekly*, 'New world for London kids', 10 February 1984.

Rob Collister: 'Adventure versus the mountain', *Adventure Education*, 1, 2, March/April 1984.

Impact of fall on tarmac: test data from Franklin Institute Research Laboratories, Philadelphia, quoted in Madders, M., *et al.*, 'Play gardens in school grounds', *BEE*, 119, March 1981.

Lin Simonon: in 'Rethinking school playgrounds', *BEE*, 119, March 1981.

Headmasters' attitudes: 'Schoolyard Landscape', op. cit.

Hilary Peters: quoted in Ward.

WCS code of ethics: CDP.

Leopold: p. 246.

Swansea tree-planter: quoted in Mostyn.

Earthways training: see Earth Education, *An Acclimatization Sourcebook*, AEI, 1981.

Chapter 10 ● EARTHWORKS

Bristol land-use project: Richards, D., 'The Sleeping Beauty in the Wood', Avon Youth Association (undated); CoEnCo/Fair Play for Children.

Bristol forestry: Bristol City Council, Woodland estate working plan (undated).

City farms: 'City roots for a green revolution', *The Times*, 26 May 1984; 'City farms on move to the urban fringe', *Farmers' Weekly*, 30 December 1983; *Soil Association Quarterly Review*, September, December 1983; 'The taming of the concrete jungle', *The Times Educational Supplement*, 5 August 1983; Wardle (1983).

Community recycling: see, for example, CoEnCo/Fair Play for Children; Lobbenberg; Juppenlatz; LEEN; *Guardian*, 7 May 1985; Conaty, P., 'An eco-view of urban development', (Birmingham) *Future Studies Centre Newsletter*, 2, 1984.

Waste-to-energy: 'Coping with waste', *The Times*, 4 February 1985.

Worthing: Thomas, C.

Glasgow: *Landscape Design*, p. 13, August 1984.

Bristol: *The Times*, 6 April 1985.

Byker: Conaty, op. cit.

Birmingham: Rufford, N., 'How rubbish became a burning issue', *Guardian*, 18 October 1984.

For developments abroad, notably Hiroshima, China (esp. Shanghai) and the Netherlands, see Chandler, W.U.; Lewis, J.W.; and Seldman.

Bottle banks: *The Times*, 4 February 1985.

Allotments: Riley.

Two million US gardeners: *Landscape*, 25, 2, 1981. See also Lewis, C.A. (1979b) and (1979c).

Corn in New York, New Jersey: Cornell University, 'Growing hope', Cooperative Extension Gardening Programme *Annual Report*, New York, 1977. See also 'Harvesting neighbourhood goodwill', *Washington Post*, 29 October 1977; Segal, S., 'The New York City gardening programme', in *Community Gardening, A Handbook* (see Lewis, C.A., 1979b).

Open space coalition: at 77 Reade St, New York 10007. See International Foundation for Development Alternatives, *Urban Self-Reliance – A Directory*, June 1985.

Bronx: Cornell University, op. cit.; Neighbourhood Open Space Coalition, Bronx Land Reclamation Program, 1982; 'No room at the bin', *The Progressive*, December 1981; Seldman and Huls (1984).

Two hundred cities recycling: *The Progressive*, op. cit.

Recycling Unlimited: Seldman (1983); Seldman and Huls.

St Paul: Seldman (1983); Seldman and Huls.

Institute for Local Self-Reliance: for further references see entries in bibliography under Morris, D., and Seldman.

Permaculture and Mollison: a good introduction is Strange. Other sources are *Permaculture*, the journal of the (Australian) National Permaculture Association, 37 Goldsmith St 2465, and *Permaculture News*, Permaculture Association, 6 Loughborough Park, London SW9 8TR.

Growing food in the city: Mollison is quoted from his address to the Soil Association of South Australia, 24 February 1983, reproduced as 'The parable of the chicken', *Permaculture*, 12 May 1983.

Compost (the Freemans): *Permaculture*, 9 April 1982.

Camden group: 'Roof garden in the smoke', *Permaculture News*, 2, Autumn 1983.

Urban forests: see Tregay (1979), Cole; Cole, L. and Mullard, J., 'Woodlands in urban areas – a resource and a refuge', *Arboricultural Journal*, 6, 4, November 1982.

Warrington: see, for example, Tregay and Gustavsson; Tregay and Moffatt; Greenwood, R.D., and Moffatt, J.D., 'Implementation techniques for more natural landscapes', Warrington and Runcorn Development Corporation (undated).

Milton Keynes: see Chapter 13.

Sweden's urban forests: *Arboricultural Journal*, October 1979.

Oakland: Beatty, R.A., 'Planning the urban forest', *Landscape Architecture*, 71, 4, July 1981.

Manchester: planting figures from Greater Manchester Council. Tree cover estimate based on Forestry Commission data. Nine million mature broadleaves would cover an area of 100,000–110,000 acres. The area of the New Forest under plantation is 30,000 acres.

Earlier New Towns: Tregay (1979).

Milton Keynes: Kelcey, J.G., 'Aspects of practical ecology', *British Ecological Society Bulletin*, 7, 1, March 1976.

Woodlands 'purify' cities: Tregay (1979); Sukopp and Werner; Walker.

Peasecroft Wood: BTCV, West Midlands Urban Conservation Project, 'Report', 1982–83.

Tree survival in Operation Greenup: Berry, P., 'The landscape, ecological and recreational evaluation of woodland', *Arboricultural Journal*, 7, 3, August 1983. Similar schemes, including free trees and shrubs for individuals and groups, are increasingly common in the UK – Liverpool, St Helens, Tower Hamlets, etc.

Bristol woodlands: op. cit.

Allotments: Riley.

CHAPTER 11 ● LARGER WORLDS: THE CITY RESHAPED

Leland: quoted in Fishwick, H., *A History of Lancashire*, Elliot Stock (undated).

Heylyn: quoted in Everitt.

Manchester's population in Defoe's time: Lampard, E.E., 'The history of cities in the economically advanced areas', pp. 81–137 in Redfield. Later figures are from Mitchell.

De Tocqueville: quoted in Briggs, p. 112.

Dereliction, pollution in Wigan, Manchester, North-West: Maund; DoE (1984).

River Croal: description from the Greater Manchester Council, 'Restoring derelict land', planning leaflet (undated).

Manchester 'dying': *The Times*, Special Report, 2 November 1984.

Manchester reclamation: the most accessible accounts are Maund; and Stuart-Murray, J., 'Bringing the countryside to the town', *Landscape Design*, 152, December 1984. See also Commons, Select Committee on the Environment, for the Greater Manchester Council Memorandum and Evidence, vol. 2, pp. 260–90.

Reclamation: DoE (1984); GMC, 'Report on 1982 Derelict Land survey', Planning committee, 16 May 1984.

Swansea comparison: see Lavender.

Urban land-use change: figures for West Midlands from the Birmingham to Wolverhampton Corridor Initiative; for Tyne and Wear from internal reports.

Newham and Sunderland: AMA.

Wembley: Wembley is 115 yards by 75 yards.

Alice Coleman: in, for example, 1985.

Mountain allegory: quotation from Lancaster, M. and Turner, T., 'The sun rises over Liverpool', *Landscape Design*, 148, April 1984.

Liverpool 'hill': see 'Mountains out of dolehills', *The Times*, 14 April 1984. Also 'Survey', *The Financial Times*, 2 May 1984.

Asphalt 'last crop': quotation attributed to M. Ruper Cutler, former US Assistant Secretary for Agriculture.

Water city: announced by London Docklands Development Corporation, 16 January 1985.

Partners for Livable Places: at 1429 21st St NW, Washington DC 20036. It has an annual newsletter, *Livability*, and a monthly journal, *Place*.

Boston and Lowell: Beioley, S., *Tourism and Urban Regeneration: Some Lessons from American Cities*, English Tourist Board, 1981.

Lowell unemployment: Clouston, B., 'A national urban heritage park', *Landscape Design*, October 1984.

British cities and tourism: see, for example, English Tourist Board, 'Tourism and the inner city', planning advisory note 3, 1980. Also ETB, *Tourism and Leisure: The New Horizon*, 1983.

Bradford: AMA; *Tourism in Action* (ETB newspaper), December 1983; 'Survey', *The Financial Times*, 8 November 1985.

Wigan: see, for example, *Guardian*, 10 September 1984.

Halifax: Jones, R.J., 'Halifax town centre conservation study', *Landscape Design*, 154, April 1985.

Manchester Ship Canal: *Guardian*, 18 February 1985.

Science parks: 'Survey', *The Financial Times*, 1 October 1984.

Henry Bennett: quoted in *The Financial Times*, 12 October 1984. See also Segal Quince and Partners, 'The Cambridge phenomenon: the growth of high-technology industry in a university town', February 1985.

Cambridge University report: *The Times*, 19 March 1985.

Swindon and Milton Keynes business parks: *Landscape Design*, 151, October 1984.

Office design: see, for example, *The Financial Times* 'Surveys' on office property, 18 May 1984, and industrial property, 2 November 1984.

Slough Estates: quotation from company advertisement.

Milton Keynes: MKDC.

Heritage centres: Centre for Environmental Interpretation, *Bulletin*, March 1985.

Local history: 'Report of the Committee to Review Local History' (Blake Report), National Council of Social Service, 1979.

Urban studies centres: CUSC.

The Ackers: West Midlands CC, 'Something for all at the Ackers'. Also *Landscape Design*, 152, December 1984; and West Midlands CC.

Leicester: the city wildlife project publishes *Leicester Wildlife News*.

Newcastle's Network: AMA. Also Tyne and Wear County Council, 'New land for old: reclaiming derelict land in Tyne and Wear' (undated); and *Riverside*, the council's six-monthly newsletter on the Tyne.

Groundwork: promoted by the Countryside Commission in conjunction with the Nature Conservancy Council and the BTCV (British Trust for Conservation Volunteers). The pilot Operation Groundwork was launched in 1981, the first five trusts in 1983. It is an extension of the urban fringe experiments of the 1970s (see above, Chapter 7).

'Villaging' the city: *Common Futures*, 2, Spring 1985.

Greenline: West Midlands CC, the Birmingham-to-Wolverhampton Corridor Initiative; also *Landscape Design*, 152, December 1984.

Rise of ecology: modern urban planning, as typified by Gordon Cullen's influential *Townscape*, has stressed the purely pictorial aspects of greenspace, unlike ecology.

Ecologists 'arrive': Elkington, J., and Roberts, J., 'Is there an ecologist in the house?', *New Scientist*, 3 November 1977.

Tulse Hill: for relative costs, see 'Schoolyard Landscape' (notes, Chapter 9).

Costs of semi-natural planting: Gilbert, O.T., 'The wildlife of Britain's wasteland', *New Scientist*, 24 March 1983; Dale, C., 'Design for maintenance and management', *Landscape Design*, August 1980; Ruff and Tregay.

Staines Moor: Davis.

Manchester ecologists: 'Mountains out of dolehills', *The Times*, 14 April 1984; Bradshaw; Bradshaw, A.D., 'The biology of land reclamation in urban areas', in Bornkamm.

Terry Wells: Wells (1981, 1983); Vines, G., 'Science can make a meadow', *New Scientist*, 18 August 1983; Gilbert, O. and Wathern, P., 'The creation of flower-rich swards on mineral workings', *Reclamation Review*, 3, 1980. For examples of public bodies using wildflower mixtures, see *Town and Country Planning*, September 1983, p. 233.

Huddersfield: Corder, M., and Brooker, R., 'Natural economy: an ecological approach to planting and management techniques in urban areas', Kirklees Metropolitan Council, 1981.

GLC plan: quoted in GLC (1984).

Glasgow: Shimwell, D.W., 'The ecological value of vacant land in the Strathclyde conurbation', a report to Strathclyde Regional Council, Manchester University, Dept of Geography, Urban Ecology Research Group, July 1983.

Dreamtime roads: Lynch (1960), p. 303.

Cardiff rivers: South Glamorgan County Council, County Planning Division, Cardiff River Valleys Recreation Study.

CHAPTER 12 • CONNECTIONS AND RECONSTRUCTIONS

Official urban aid and MSC special programme spending: DoE, 'Urban Programme Resources'; DoE, 'Inner Cities Directorate', 'Inner-city policy, England', February 1984; MSC Annual Reports, 1975–84. Expenditure in cash prices. For further explanation, see note below.

Job creation framework: Commons Select Committee on Expenditure. Also Hedges, B. and Courtenay, G., MSC 'Special programmes sponsors survey, Social and Community Planning Research 1983.

Diffusion of power: the figure for MSC spending (£3.3 billion) includes job creation programme, community industry, STEPS, YOPS, and community and voluntary projects programmes but excludes work experience programmes and the youth training scheme; the latter are more specifically 'vocational'. It is likely that the ratio of 1.35:3.3 substantially understates the MSC's greater role in enhancing the local 'unofficial' sector. 'Traditional' (voluntary) urban programme projects are appraised centrally, unlike MSC projects. Much 'blurring' of divisions between different MSC schemes has occurred in response to social needs (Commons Select Committee on Expenditure, vol. 2, p. 51). In 1983–84, of £310 million urban programme spending, £62 million was spent on voluntary projects: the MSC meanwhile spent £429 million on the community and voluntary projects programmes and on community industry (see ibid., vol. 2, p. 131). On precedents for the diffusion of power, three suggest themselves: aid to industry (under the 1972 Industry Act, for example); grants to established voluntary bodies (usually their headquarters); and transfer payments (for example, social security) to private individuals. The job creation model differs radically from them all.

Work ethic: on its decline, and weakness in Britain, see 'Human values in working life', *Employment Gazette*, February 1984.

'Community' and its loss: Berry (1973); Bell and Newby.

Burgess: see Park *et al.*

Home-centredness: see, for example, Henley Centre for Forecasting, 'Leisure Futures', Autumn 1984.

Self-employment: from 1.84 to 2.5 million between 1979 and 1984 (Manpower Services Commission, Labour Market Quarterly Reports).

Homeworking: see *Employment Gazette*, January 1984. Home and freelance working grew from 1.1 million in the late 1960s to 1.7 million.

DIY: the market grew in real terms by a third between 1977 and 1983. In the decade 1977–87 projected growth is 47 per cent (Henley Centre for Forecasting, op. cit.).

Employment accommodation: 'UK property survey', *The Financial Times*, 30 November 1984.

Factories' in farms: see, for example, 'Small businesses survey', *The Financial Times*, 12 June 1984, p. 13, on work of the Council for Small Industries in Rural Areas (COSIRA).

Revaluation of work: Handy; Robertson; Blandy, A., 'New technology and flexible patterns of working time', *Employment Gazette*, October 1984.

Use-classes obsolete: see survey by Incorporated Society of Valuers and Auctioneers in 1984; 'Industrial property survey', *The Financial Times*, 2 November 1984.

Bell and Puritan ethic: Bell (1976).

Localism and city expansion: Young and Garside; Aston and Bond; Johnson. For modern parallels and lessons, see Bar-Gal, Y., 'Villages under siege', *Ekistics*, 5, 298, 1983.

Urban-rural contrasts: the debate is partly covered in Berry (1973). For case studies, see Harrison and Gibson; Carter, pp. 356–60.

Crowding, non-persons and coping strategies: the subject is well covered in *The Psychology of Urban Life*, special issue of *Environment and Behaviour*, 10, 2, June 1978. Quotations from various articles. See also Berry (1973); McLelland, L., and Auslander, N., 'Perceptions of crowding and unpleasantness in public settings', *Environment and Behaviour*, 10, 4, 1978.

Systems, organization theory: Pugh; Laszlo.

Abaluyia: Weisner, T., 'Recurrent migration and rural-urban differences in Kenya children's social behaviour', paper presented at Psychosocial Conference of Sedentarization, University of California at Los Angeles, 1974.

Ecological psychology: Gump, P.V. and Adelberg, B., 'Urbanism from the perspective of ecological psychologists', *Environment and Behaviour*, 10, 2, 1978. Quotations: ibid. See also Barker (1968).

Detail in mental maps: Lee, T.R., 'Psychology and living space'; Stea, D. and Blaut, J.M., 'Some preliminary observations on spatial learning in schoolchildren'; both in Downs and Stea; Van Vliet, W., 'Children's travel behaviour', *Ekistics*, 50, 298, 1983.

Children cycling and self-reliance: see, for example, Ward. The study quoted is from Sully, J., 'Child cycling', *Town and Country Planning*, June 1976. The desire to 'reconnect' with surroundings may also help to explain the growth in cycle traffic – of 59 per cent, nationally, 1973–83 and 81 per cent, 1976–83, in London (figures: Department of Transport, GLC).

'Seeing' and 'learning': Gump and Adelberg, op. cit.

Lynch: quoted in Ward.

Psychological health predictors: Antonovsky.

'New localism': Morris and Hess. For a criticism of the 'moral sacredness' of local territory, see Sennett (1977).

Environmental localism: Lowe and Goyder; Social Trends (1985), Central Statistical Office/HMSO.

Local economy: see, for example, Cobbold, C., 'Working together for jobs', *Town and Country Planning*, October 1983; Coffey, W. and Polese, M., 'Local development: conceptual bases and policy implications', *Regional Studies*, 19, 2, April 1985; Allan, M., 'Local employment strategies', *The Planner*, 69, 5, September-October 1983; Windass, S. (ed.), *Local Initiatives in Great Britain*, Foundation for Alternatives, Oxford, 1981; 'The big guns backing Mr. Small', *The Times*, 16 October 1984; 'Generating jobs', *The Times*, special report on the CBI special programmes unit, 24 February 1984; Ekins; Martin and Hodge.

Branch-plant economy: Wray; Gaffikin and Nickson.

Self-managed economy: Worker co-operatives grew from under 100 in 1977 to over 900 in 1984 (Thomas, D., 'Working together', *New Society*, 27 September 1984). See above for self-employment. See also 'Co-operatives and local economic development', *Regional Studies*, 17, 4, 1983. On community businesses and workspace projects see, for example, Pearce, J. and Hopwood, S., 'Create your own jobs – community-based initiatives', *The Planner*, 67, 3, May-June 1981; 'Where charity and profit meet', *The Financial Times*, 9 July 1985.

Local enterprise agencies: a thought-provoking article is O'Brien, S., 'Alfonzo's challenge to social entrepreneurs', *The Financial Times*, 13 November 1985.

Self-build: figure from *The Times*, 12 December 1984. See also *Housing Review*, January-February 1984.

Outdoor education: see, for example, 'Outdoor education and the urban child', Report of NAOE's annual conference, 1981; 'Towards outdoor education', Report by National Centre for Advice and Information on Outdoor Activities, Doncaster (undated).

Community organizers: *Self-Reliance* (ILBR newsletter), 24 September-October 1980.

Technical aid centres: Wates, N., 'Actac in action', *Architects Journal*,

12 October 1983; various issues of *Community Technical Aid* (ACTAC newsletter).

Birkenhead and planning for real: Gibson; Jubilee Enterprise Trust, 'Making the most of local resources, self-help feasibility study of the Conway area of Birkenhead', 1983; Gibson, A., 'Planning for real', *Voluntary Action*, Summer 1980; Duckenfield, M., 'Planning for real', *New Society*, 2 August 1979. Village appraisals are a related rural initiative: see Clark, D., 'Community initiatives in the countryside', *The Planner*, 72, 2, February 1986.

Wilderness therapy: the literature is mainly American. See pp. 99–108 of Pepper, B. and Ryglewicz, H. (eds), 'Advances in treating the young adult chronic patient', *New Directions for Mental Health Series*, no. 21, Jossey-Bass, March 1984; Turner, A.L., 'The therapeutic value of nature', *Journal of Operational Psychiatry*, 7, 1, 1976; Kraus, I. W., 'The effectiveness of wilderness therapy with emotionally disturbed adolescents', *Dissertation Abstracts International*, 44, 5B, November 1983; Bernstein, A., 'Wilderness as a therapeutic behaviour setting', *Therapeutic Recreation Journal*, 6, 1972.

Horticultural therapy: Woodward, S., 'Horticulture – is it therapy?' *Growth Point*, 17, Autumn 1985; Society for Horticultural Therapy, 'Opportunities in Environmental Improvement', November 1983; Leonhardt, G., 'Reflections of a city gardener', *Landscape*, 26, 2, 1982; Lewis (1979a and b, 1983); Kaplan (1973, 1983).

Gardeners' survey: Lewis (1979c, 1983).

Environment, jobs and entrepreneurs: see, for example, NCC; Lewis (1979a); O'Brien, op. cit.

'Neutral' space: see Rapoport (1980). The charity Common Ground regards its parish maps project as community therapy: see Deakin, R., 'The nature of "common ground"', *Common Futures*, 3, Summer 1985.

Tulse Hill: see the Tulse Hill Nature Garden (from the Estate Office, 13 Purser House, London SW2).

Weller Street: Gibson.

Birkenhead: quotations from Jubilee Enterprise Trust, op. cit.

Greenspace and neighbourhood boundaries: DoE (1974).

Tree survival: The 20–25 per cent survival rate, for example, is the figure for Sunderland, quoted at the NAOE's conference, 'Outdoor Education and the Urban Child', op. cit.

Liverpool trees: by 1981 replacements for Operation Eyesore were themselves being replaced.

Junior schools and trees: 'Schoolyard landscape', op. cit. (Chapter 9), p. 4. See also *Bulletin of Environmental Education*, 108, April 1980.

Arnstein: see, for example, Ward and Fyson.

Thirty-seven council officers: Collins, T. (Oasis Children's Venture), 'All we needed was permission', *Better Times*, 3, 1983.

Nicholson: in 'Life in the city', *Naturopa*, 36, 1980.

Public involvement schemes: Community Projects Foundation, 'Public Involvement in Local Government: A Survey in England and Wales', 1985.

Urban parishes: see the *Guardian*, p. 13, 21 August 1985.

London Wildlife Trust: on its fifth anniversary in 1986 the Trust had 3,000 members and thirty local groups.

Naturalists' trusts: *Natural World*, Winter 1985. For further implications on the economics of autonomy, see Brooker, R., 'Providing a maximum quality environment at minimum cost', *Municipal Journal*, 87, 42, 1979.

Rollestone Wood: Tartaglia-Kershaw, M., 'The recreational and aesthetic significance of urban woodland', *Landscape Research*, 7, 3, Winter 1982.

CHAPTER 13 • BEYOND THE CITY

Oxfam: *Observer*, 5 February 1984.

UNEP and soil erosion: *New Scientist*, 10 May 1984, p. 3.

'Natural' disasters increasing: Earthscan.

Third World city: see for example, Hardoy and Satterthwaite; Blitzer *et al.*; *World Health* (special issue), 'In the shadow of the city', July 1983; McAuslan.

Urban bias: Lipton; O'Connor; Gillie; *Guardian*, 'The false starts that Africa failed to correct', 8 April 1985; Kulaba, S.M., 'Rural settlement policies in Tanzania', *Habitat International*, 6, 1/2, 1982; Brebner, P., and Briggs, J., 'Rural settlement planning in Algeria and Tanzania: a comparative study', *Habitat International*, 6, 5/6, 1982.

Electricity consumption: Agarwal *et al.*, p. 38.

World urbanization and land: in 1980 the level of Third World urbanization was 41 per cent (Eckholm, p. 28) compared with western levels of 80–90 per cent.

Estimate of eleven billion: World Bank. The calculation is an 'intermediate' projection based on 'high' American car ownership levels (one car for two people: Brown, L.R.) but 'low' English road usage (24 yards per car: CSO) and road width (20 feet) standards. It is for road tarmac alone and excludes all associated land take or blight (pavements, reservations, etc.). 20 feet is a typical local road width. The smallest trunk road is 24 feet wide (Department of Transport).

Aid as contagion: Hanlon, J., 'Tanzania decides how to industrialize', *New Scientist*, 21 September 1978; 'Aid grows a crop of problems', *Guardian*, 2 December 1983; Toye, J., 'The case for hard cash in aid for the Third World', *The Times*, 24 May 1984; Cassen, R., 'Does aid really work?', *New Society*, 26 July 1985.

Himalayan forestry: *New Scientist*, 4 November 1982.

Green belt movement: International Foundation for Development Alternatives, Dossier 49, September-October 1985; TOES.

Power dispersal, etc.: Naisbitt; Toffler.

Countryside values: Two-fifths of the population of England and Wales visit the countryside on a typical summer Sunday. Eighty-five per cent visit it during the year (Countryside Commission, 1985). See also survey in Hall *et al.* (Chapter 1).

Environmental quality: for a modern version of a future unpolluted London, see *Town and Country Planning*, July-August 1983.

Boundaries and settlement: Pennick.

Urban demolition: CPF (1983). See also Wray.

City housing and economic dependency: see Elias and Keogh for penalties of reverse commuting. Also Wray.

Urban densities: Best.

Space and house moves: Clark and Onaka; Hall *et al.*

'Ideal' dwellings etc.: Marcus and Sarkissian.

'Harmony with nature': Petersen, G.L., 'A model of preference: quantitative analysis of the perception of the visual appearance of residential neighbourhoods', *Journal of the Regional Science Association*, 7, 1967.

Density-size rule: Best, p. 69.

Pivotal density: Best, p. 77; Hall *et al.*, vol. 2, p. 299.

Housing needs: figures from NFHA.

City percentages: figures from Housing Research Foundation, Housing and land, 1984–91: 1992–2000, Paper 3, 1984; and memorandum by House-Builders' Federation to Commons (1984), 'Urban land supply and green belt'. The HBF put the absolute maximum at 15 per cent but said 10 per cent was likelier.

Cobbett: In *Rural Rides*.

Diet and health: see Vogtmann.

Chinese peasant: Leach. See also Smit, J., 'Urban and metropolitan agricultural prospects', *Habitat International*, 5, 3/4, pp. 499–506, 1980; Maiken.

Diet and land: see Williams, W., 'UK food production: resources and alternatives', *New Scientist*, 8 December 1977; Mellanby; Waller, R., 'The agricultural balance sheet', Green Alliance/Conservation Society, 1982. Such calculations are bedevilled by different assumptions. The essential point is that the land-take resulting from a non-meat diet is lower by a factor of two or three.

Two per cent of land needed: Strange.

Russia: ibid.

NFU, CLA: see, for example, Country Landowners' Association, 'Maintaining income from land', November 1985. Quotation is from Simon Gourlay, then deputy NFU president, at the 1986 Oxford farming conference.

Lightmoor: 'Brave new garden city', *The Times*, 4 July 1984.

Communes, etc.: see, for example, McLaughlin and Davidson; Rigby; Communes Network, *The Collective Experience*, Leicester, 1984. Some accounts are: *Findhorn*: 'The big guns backing Mr Small', *The Times*, 16 October 1984. *Scoraig*: *Guardian*, 3 November 1984.

Tintern, Greentown: Gibson, pp. 49, 130. Greentown plan to use the design methods advocated by Christopher Alexander. A venture similar to Lightmoor is the Pentref Development Trust's 'new village' of 500 homes in the Penllergaer Forest near Swansea.

Integrated rural development: *Landscape Research*, 10, 2, Summer 1985. Also MacEwan, M., 'A future for national parks', *Ecos*, 1, 2, Spring 1980.

New towns: see Turner, T., 'Planning the landscape for a new town', *Town and Country Planning*, November 1982.

Redditch: Laurie (1979); Brown, J.; Turner, op. cit.; *Landscape Research*, 6, 3, Winter 1981.

Telford: Walker, C., and Tobin, R.W., 'Telford new town', *Naturopa*, 36, 1980.

New Ash Green: Clouston and Stansfield, p. 137.

Warrington: Ruff and Tregay; Tregay and Gustavsson; Tregay and Moffatt; Greenwood, R., 'Gorse Covert, Warrington – creating a more natural landscape', *Landscape Research*, 143, June 1983; *The Planner*, November-December 1982; *Bulletin of Environmental Education*, Special issue, 'Education in Urban Parks', 139, December 1982; Tregay, R., 'Community involvement – or marketing the product?', *Landscape Design*, 155, June 1985.

Birchwood Science Park: *The Planner*, November-December 1982.

Milton Keynes: Walker, D.; MKDC; Kelcey, J.G., 'Ecology and development', MKDC, 1977; Yoxon, M., 'Environmental education in Milton Keynes', MKDC, 1981; Kelcey, J.G., 'Opportunities for wildlife on road verges in a new city', *Urban Ecology*, 1, 1975; Kelcey, J.G., 'Ecology in a new British city', *Naturopa*, 22, 1975; Kelcey, J.G., 'Aspects of practical ecology', *British Ecological Society Bulletin*, VII.1, March, 1976; Harris, M., 'The houses we choose', *New Society*, 13 December 1984.

Cyclists and pedestrians: *The Times*, Special Report, 29 October 1984. *Bicycles as appropriate technology*: in 1982 the Intermediate Technology Development Group organized seminars to 'explain' the bicycle to local authorities.

Self-build: *The Times*, Special Report, 29 October 1984.

Energy and settlement patterns: Owens, S., 'Spatial structure and energy demand', in Cope *et al.*

Europe 2000: Hall (1977).

Cambridge: see, for instance, Cambridgeshire County Council.

Creative ecology: see Bradshaw; Bradshaw and Chadwick; Wells, T.C.E. (1981, 1983); Grime; Kelcey; Ratcliffe.

Rother Valley: Rother Valley Country Park Joint Committee, 'The making of the Rother Valley Country Park', 1983; Bannister, A. and Moorhouse, P., 'The making of Rother Valley country park', *Landscape Design*, 152, December 1984; South Yorkshire County Council, 'Ecological work', Environment department information report (undated).

Ecodevelopment body: UK Centre for Economic and Environmental Development.

Bioshelter: Todd and Todd. Also Cornish.

Biotecture: Doernach, R., 'On the use of biotectural systems', *Permaculture*, 7, March 1981.

Oekotop: *Permaculture*, 9, April 1982.

Planning for wildlife: Sukopp and Werner; Laurie (1979), (Chapter 8); Bornkamm; Moore, P., 'The ecology of diversity', *New Scientist*, 28 February 1985; Goldstein, E.L., Gross, M., and Marston, A.L., 'A biogeographic approach to the design of greenspace', *Landscape Research*, 10, 1, Spring 1985.

Poland: Luniak, M., 'An experiment in Poland', *Naturopa*, 36, 1980.

American study: by the University of Arizona. See Ittelson, W., 'Environmental perception and urban experience', *Environment and Behaviour*, 10, 2, June 1978.

Ashland: Mudrak.

Rural settlement and wildlife: see, for example, Mills, S., 'French farming: good for people, good for wildlife', *New Scientist*, 24 November 1983. On farms versus housing for wildlife, see Davis; Ratcliffe.

Consortium developments: the plans were announced in 1984. The Tillingham Hall inquiry was in 1986.

Quantity and quality: a landmark in the improvement of housing landscape design was the Essex Design Guide, published in 1973 and increasingly adopted by local authorities.

Walnut trees: Strange.

Mollison: estimate in 'The parable of the chicken', *Permaculture*, 12, May 1983. John Jeavons's estimate of the land required for a complete diet for one person using bio-intensive methods is 25 feet by 28 feet over an eight-month growing season (Maiken).

British garden size: Best, p. 107.

Gardens controversy: Best. The calculation that follows assumes an (average) household size of 2.64, an urban area of 4.7 million acres and a net residential component of 49 per cent (figures, for England and Wales, based on Best and NFHA). Since many households do not require a garden, the cut in urban area would in practice be much greater.

Chemical-free food: Vogtmann.

Land-use, farming and urban densities: Best.

Urban areas: Best; Housing Research Foundation, op. cit.

Green belts: the case against green belts is well summarized in Commons Select Committee on the Environment by Willis, K.G. and Whitby, M.C. (Appendix 11, see especially pp. 586–8).

'Brown' land: Munton, R.J.C., ibid. (Appendix 9, p. 565). See also Evans, A.W., 'The economic consequences of green belts', Memorandum, Commons Select Committee on the Environment; Fitton, M., 'The urban fringe and the less privileged', *Countryside Recreation Review*, 1, 1976; Harrison, C., 'Countryside recreation and London's urban fringe', *Transactions of the Institute of British Geographers*, 8, 1983; Elson, M.J., 'The use of green belts – conflict mediation in the pressured countryside', *The Planner*, 72, 2, February 1986; Shoard (1979).

Dispersal and population growth: Hall *et al.*

Odum: Odum, E.P. and Odum, H.T., 'Natural areas as necessary components of man's total environment', *Transactions of the 37th North American Wildlife and Natural Resources Conference* (Mexico City), Wildlife Management Institute, 1972.

Medieval town: Braudel.

Bronx: *Landscape*, 27, 2, 1983.

New towns: the case for a second programme is argued, for example, in Wray, I., 'Can the South boom again?', *Architects' Journal*, 19 June 1985.

Chinese cities: Lewis, J.W.; Salter, C.L., 'China shapes the city with bold strokes', *Landscape Architecture*, 70, 1, January 1980; Ravetz.

Divine perspective: quotation from Cicero's *De Natura Deorum*. See also Russell, P.

Wolf: Mills, S., 'The big bad wolf', *New Scientist*, 29 November 1984.

Tiger: 'Saving India's Wildlife', *The Futurist*, 20, 1, January-February 1986.

Zimbabwe: du Toit, R., 'A middle way for wildlife parks', *New Scientist*, 31 January 1985.

BIBLIOGRAPHY

Adams, R., *Watership Down*, Penguin, 1974.

Adler, M., *Drawing Down the Moon: The Resurgence of Paganism in America*, Viking, 1979.

Agarwal, A., Satterthwaite, D. and Lean, G., *Life at the Margin: The Need for Third World Urban Development*, Earthscan, 1979.

Allen, M. (Lady Allen of Hurtwood), *Planning for Play*, Thames & Hudson, 1960.

Alonso, W., *Location and Land Use*, Harvard University Press, 1964.

AMA (Association of Metropolitan Authorities), 'Green Policy', 1985.

Antonovsky, A., *Health, Stress and Coping*, Jossey-Bass, 1979.

Appleton, Jay, *The Experience of Landscape*, John Wiley, 1975.

Aston, M., and Bond, J., *The Landscape of Towns*, Dent, 1976.

Bacon, E., *Design of Cities*, Thames & Hudson, 1967.

Bagrow, L.A., *A History of Cartography*, C.A. Watts, 1964.

Baines, C., *How to Make a Wildlife Garden*, Hamish Hamilton (Elm Tree Books), 1985.

Barker, R.G. (ed.), *Ecological Psychology*, Stanford University Press, 1968.

Barr, John, *Derelict Britain*, Penguin, 1969.

Becker, L.C., *Property Rights: Philosophic Foundations*, Routledge & Kegan Paul, 1977.

Bell, C. and Newby, H., *The Sociology of Community*, Frank Cass, 1974.

Bell, D., *The Coming of Post-Industrial Society*, Heinemann, 1974.

Bell, D., *The Cultural Contradictions of Capitalism*, Heinemann, 1976.

Bell, G. and Tyrwhitt, J. (eds), *Human Identity in the Urban Environment*, Penguin, 1972.

Benevolo, L., *The History of the City*, Scolar Press, 1980.

Bentham, C.G., 'Urban problems and public dissatisfaction in the metropolitan areas of England', *Regional Studies*, 17, 5, 1983.

Berger, J., *Ways of Seeing*, BBC, 1972.

Bergon, Frank (ed.), *The Wilderness Reader*, New American Library, 1980.

Berry, B.J.L., *The Human Consequences of Urbanization*, Macmillan, 1973.

Berry, B.J.L., *Urbanization and Counter-Urbanization*, Sage, 1976.

Berry, D., 'Landscape aesthetics and environmental planning: a critique of underlying premises', *Regional Science Research Institute Paper* 85, Philadelphia, 1975.

Best, R.H., *Land Use and Living Space*, Methuen, 1981.

Bilksky, C.J., (ed.), *Historical Ecology: Essays on Environmental and Social Change*, 1980.

Blitzer, S., Hardoy, J.E. and Satterthwaite, D., *Habitat – Five Years After*, Earthscan, 1981.

Bornkamm, J. (ed.), *Urban Ecology*, Blackwell, 1982.

Bradshaw, A.D., 'Conservation problems in the future', *Proceedings of the Royal Society of London*, 8, 197, 1977.

Bradshaw, A.D. and Chadwick, M.D., *The Restoration of Land*, Blackwell, 1980.

Braudel, F., *Capitalism and Material Life, 1400–1800*, transl. M. Kochan, Weidenfeld & Nicolson, 1973.

Briggs, A., *Victorian Cities*, Odhams, 1963.

Brown, J., *The Everywhere Landscape*, Wildwood House, 1982.

Brown, L.R. *Building a Sustainable Society*, W.W. Norton, 1981.

BTCV (British Trust for Conservation Volunteers), Conservation Project Park, 1980.

Burgess, J.A., 'Selling place: environmental images for the executive', *Regional Studies*, 16, 1, 1982.

Burnett, J., *A History of the Cost of Living*, Penguin, 1969.

Burrows, J., 'Vacant urban land – a continuing crisis', *The Planner*, 64, 7–9, 1978.

Butler, D. and Sloman, A., *British Political Facts, 1900–1979*, Macmillan, 1979.

Cambridgeshire County Council, *Structure Plan*, 1980, *Annual Monitoring Reports*, 1981 and 1982.

Cameron, G. (ed.), *The Future of the British Conurbation*, Longman, 1980.

Capra, F., *The Turning Point: Science, Society and the Rising Culture*, Wildwood House, 1982.

Carpenter, H., *Tolkien: A Biography*, Allen & Unwin, 1977.

Carpenter, H., *Secret Gardens: A Study of the Golden Age of Children's Literature*, Allen & Unwin, 1985.

Carter, H., *The Study of Urban Geography*, 3rd edn., Edward Arnold, 1981.

CDP (Conservation and Development Programme for the UK), *A Response to the World Conservation Strategy*, Kogan Page, 1983.

Chadwick, G.F., *The Park and the Town*, Architectural Press, 1966.

Champion, A.G., 'Population trends in rural Britain', *Population Trends*, 26, Winter, 1981.

Chandler, J.J., *The Climate of London*, Hutchinson, 1965.

Chandler, W.U., *Materials Recycling: The Virtue of Necessity*, Paper 56, Worldwatch Institute, 1983.

Cherry, G.E., *Urban Change and Planning: History of Urban Development in Britain since 1750*, Foulis, 1972.

Cherry, G.E., *The Evolution of British Town Planning*, Halsted, 1974.

Chinery, M., *The Natural History of the Garden*, Collins, 1977.

Chinery, M., *The Living Garden: A Practical Guide to Attracting and Conserving Garden Wildlife*, Dorling Kindersley, 1986.

Chisholm, M., 'City, region and – what kind of problem?', in Patten, J. (ed.), *The Expanding City*, Academic Press, 1983.

Chorley, R., *Atmosphere, Weather and Climates*, Methuen, 1982.

Civic Trust, *Urban Wasteland*, Civic Trust, 1977.

Clark, W.A.V. and Onaka, J.L., 'Life cycle and housing adjustment as explanations of residential mobility', *Urban Studies*, 20, 3, 1983.

Clouston, B. and Stansfield, K. (eds), *Trees in Towns*, Architectural Press, 1981.

CoEnCo/Fair Play for Children, *Waking Up Dormant Land*, CoEnCo, 1981.

Cole, L., 'Wildlife in the city: a study of practical conservation projects', Nature Conservancy Council, 1980.

Coleman, A., 'The death of the city', *The London Journal*, 6, 1, 1980.

Coleman, A., *Utopia on Trial: Vision and Reality in Planned Housing*, Shipman, 1985.

Collingwood, R.G., *The Idea of Nature*, Clarendon Press, 1945.

Commons Select Committee on Expenditure, Seventh Report, 1976–77, *The Jobs Creation Programme*, vols I (*Report*) and II (*Minutes of Evidence*), HMSO, 1977.

Commons Select Committee on the Environment, First Report, 1983–84, *Green Belt and Land for Housing*, vols I (*Report*), II (*Minutes of Evidence*) and III (*Appendices*), HMSO, 1984.

Cope, D.R., Hills, R. and James, P., *Energy Policy and Land-Use Planning*, Pergamon, 1984.

Cornell, J., *Sharing Nature With Children*, Exley/Inter-Action Inprint, 1982.

Cornish, E. (ed.), *Habitats Tomorrow: Homes and Communities in an Exciting New Era*, World Future Society, 1983.

Countryside Commission, *The Bollin Valley – A Study of Land Management in the Urban Fringe*, Countryside Commission, 1976.

Countryside Commission, *1984 Countryside Recreation Survey*, Countryside Commission, 1985.

CPF (Community Projects Foundation), *Community Development – Towards a National Perspective*, CPF, 1982.

CPF, *Urban Renewal: Securing Community Involvement*, CPF, 1983.

Crone, J.R., *Maps and their Makers*, Hutchinson, 1966.

CSO (Central Statistical Office), Annual Abstract of Statistics, HMSO, 1985.

Cullen, G., *Townscape*, Architectural Press, 1961.

CUSC (Council for Urban Studies Centres), *State of the Art Report*, Town and Country Planning Association/Streetwork, 1984.

Dahmann, D.C., 'Subjective assessment of neighbourhood quality by size of place', *Urban Studies*, 20, 1, pp. 31–45, 1983.

Daniken, Erich von, *Chariots of the Gods?*, Souvenir Press, 1969.

Darling, F.F., *Wilderness and Plenty*, Ballantyne, 1970.

Davies, P., *God and the New Physics*, Dent, 1983.

Davis, B.N.K., 'Wildlife, urbanization and industry', *Biological Conservation*, 10, 1976.

Defoe, D., *A Tour Through England and Wales, 1724–27*, Penguin, 1979.

Dennis, R., 'The decline of manufacturing employment in Greater London: 1966–74', *Urban Studies*, 15, 63–73, 1978.

DoE (Department of the Environment), Neighbourhood Councils in England, consultation paper, HMSO, 1974.

DoE, *Inner Area Studies: Liverpool, Birmingham and Lambeth*, HMSO, 1977.

DoE, *Survey of Derelict Land in England, 1982*, HMSO, 1984.

Douglas, I., 'The city as an ecosystem', *Progress in Physical Geography*, 5, 3, 1981.

Downs, R.M. and Stea, D. (eds), *Image and Environment*, Aldine Publishing, 1973.

Doxiadis, C.A., *Ecology and Ekistics*, ed. G Dix, Elek, 1977.

Duncan, J.S. (ed.), *Housing and Identity: Cross-Cultural Perspectives*, Croom Helm, 1981.

Dyos, H.J., *Exploring the Urban Past: Essays in Urban History*, Cambridge University Press, 1982.

Dyos, H.J. and Woolfe, M., *The Victorian City: Images and Realities*, Routledge & Kegan Paul, 1973.

Earthscan, *Natural Disasters: Acts of God or Acts of Man?*, Earthscan, 1984.

Eckholm, E.P., *Down to Earth*, Pluto Press, 1982.

Eiseley, Loren, *The Immense Journey*, Random House, 1967.

Ekins, P. (ed.), *The Living Economy: A New Economics in the Making* (Proceedings of The Other Economic Summit, 1984–85), Routledge & Kegan Paul, 1986.

Eliade, M., *Shamanism*, Routledge & Kegan Paul, 1964.

Eliade, M., *Myths, Dreams and Mysteries*, Collins (Fontana), 1968.

Eliade, M., *Patterns in Comparative Religion*, Stagbooks, 1979.

Eliade, M., *Cosmos and History*, Routledge & Kegan Paul, 1982.

Elias, P., and Keogh, G., 'Industrial decline and unemployment in the inner city areas of Great Britain: A review of the evidence', *Urban Studies*, 19, 1, 1982.

Everitt, A., 'The market towns', in Thirsk, J. (ed.), *The Agrarian History of England and Wales*, vol. 4, Cambridge University Press, 1967.

Fairbrother, N., *New Lives, New Landscapes*, Architectural Press, 1970.

Fielding, A.J., *Counterurbanization in Western Europe, Progress in Planning*, vol. 17, Pergamon, 1982.

Fisher, F.J., 'London as an engine of economic growth', in Bromley, J.S. and Kossmann, E.H. (eds), *Britain and the Netherlands*, vol. 4, Martinus Nijhoff, 1971.

Fitter, R.S.R., *London's Natural History*, Collins, 1945.

Fordham, F., *An Introduction to Jung's Psychology*, Penguin, 1953.

Foster, C.D. and Richardson, R., 'Employment trends in London in the 1960s and their relevance for the future', in Donnison, D. and Eversley, D. (eds), *London: Urban Patterns, Problems and Policies*, Heinemann, 1973.

Fothergill, S. and Gudgin, G., *The Job Generation Process in Britain*, Centre for Environmental Studies, London, 1979.

Fothergill, S., *et al.*, *Industrial Location Research Project*, Working Papers nos. 1, 2, 3, 4 and 6, University of Cambridge, Dept of Land Economy, 1982–83.

Fothergill, S., *et al.*, 'The impact of the new and expanded town programmes on industrial location', *Regional Studies*, 17, 4, 1983.

Gaffikin, F. and Nickson, A., *Jobs Crisis and the Multinationals: The Case of the West Midlands*, Third World Publications, 1984.

Geddes, P., *Cities in Evolution*, 1915, repr. Fertig, 1969.

Gershuny, J., *After Industrial Society: The Emerging Self-Service Economy*, Macmillan, 1978.

Gibson, A., *Counterweight: The Neighbourhood Option*, Town and Country Planning Association/Education for Neighbourhood Change, 1984.

Gillie, P., 'Planning law and land tenure in developing countries', *Habitat International*, 4, 4–6, 1979.

Glacken, C.J., *Traces on the Rhodian Shore (Nature and Culture in Western Thought from Ancient Times to the End of the Eighteenth Century)*, University of California Press, 1967.

GLC (Greater London Council), 'Economic statistics from the 1981 census', 1981.

GLC, 'Ecology and nature conservation in London', 1984.

GLC, 'London's industrial strategy', Introduction, 1985.

Gold, J. and Burgess, J. (eds), *Valued Environments*, Allen & Unwin, 1982.

Gottmann, J., *Megalopolis*, Twentieth Century Fund, 1961.

Gould, P. and White, R., *Mental Maps*, Penguin, 1974.

Graber, Linda H., *Wilderness as Sacred Space*, Association of American Geographers, 1976.

Gregory, D., *Green Belts and Development Control*, Centre for Urban and Regional Studies, University of Birmingham, 1970.

Grime, J.P., 'The creative approach to nature conservation', in Ebbing, F.J. and Heath, G.W., 'The future of man', *Institute of Biology Symposium*, 20, 1972.

Gudgin, G., *Industrial Location Processes and Regional Employment Growth*, Saxon House, 1978.

Gudgin, G., Moore, B. and Rhodes, J., 'Employment problems in the cities and regions of the UK: prospects for the 1980s', *Cambridge Economic Policy Review*, 8, 2, 1982.

Hadfield, M., *A History of British Gardening*, John Murray, 1979.

Hall, P., *The World Cities*, Weidenfeld & Nicolson, 1984.

Hall, P. (ed.), *Europe 2000*, Duckworth, 1977.

Hall, P. and Hay, D., *Growth Centres in the European Urban System*, Heinemann, 1981.

Hall, P., *et al.*, *The Containment of Urban England*, Allen & Unwin, 1973.

Hallett, G., *Second Thoughts on Regional Policy*, Centre for Policy Studies, 1981.

Hammett, W.E., 'Cognitive dimensions of wilderness solitude', *Environment and Behaviour*, 14, 4, 1982.

Hamnett, C., 'The lost gentrifiers', *New Society*, 15 March 1984.

Handy, C., *The Future of Work*, Blackwell, 1984.

Hardoy, J.E. and Satterthwaite, D., 'Third World cities and the environment of poverty', *Geoforum*, Summer, 1984.

Hardy, A., *The Biology of God*, Cape, 1975.

Hardy, D. and Ward, G., *Arcadia for All: The Legacy of a Makeshift Landscape*, Mansell, 1984.

Harrison, G.A. and Gibson, J.B. (eds), *Man in Urban Environments*, Oxford University Press, 1976.

Hart, R.A. and Moore, G.T., 'The development of spatial cognition: a review', in Downs and Stea, op. cit.

Harvey, D., *Social Justice and the City*, Edward Arnold, 1973.

Hauser, P.M. and Gardner, R.W., 'Urban future: trends and prospects', paper presented at International Conference on Population and the Urban Future, Rome, September 1980.

Hausner, V. and Robson, B., *Changing Cities*, ESRC, 1985.

Hawley, A.J., *Human Ecology*, Wiley, 1950.

Hodge, G., 'Canadian small town renaissance: implication for settlement system concepts', *Regional Studies*, 17, 1, 1983.

Hoskins, W.G, *Provincial England*, Macmillan, 1965.

Howard, Ebenezer, *Garden Cities of Tomorrow* (1902), Faber, 1970.

Hughes, J.D., *Ecology in Ancient Civilization*, University of New Mexico Press, 1975.

Hunt, J.D. and Willis, P., *The Genius of the Place: The English Landscape Garden 1620–1820*, Paul Elek, 1975.

Hunter, J., *Land Into Landscape*, George Godwin, 1985.

Illich, I. *et al.*, *Disabling Professions*, Marion Boyars, 1977.

Inglehart, R., *The Silent Revolution: Changing Values and Political Styles among Western Publics*, Princeton University Press, 1977.

Institute for Local Self-Reliance, Background paper, 'Recycling Research Agenda Conference for National Science Foundation, October 1979.

Institution of Civil Engineers, *First Report of the Infrastructure Planning Group*, ICE, 1984.

International Labour Office, 'Employment and large cities: problems and outlook', 1983.

Jacobs, J., *The Economy of Cities*, Random House, 1969.

Jefferies, Richard, *After London*, Cassell, 1885.

Johnson, J. (ed.), *Suburban Growth – Geographical Processes of the Edge of the City*, Wiley, 1974.

Jones, Lang, Wootton, 'The decentralization of offices from central London', JLW Research, June 1983.

Jung, C.G., *Modern Man in Search of a Soul*, Routledge & Kegan Paul, 1933.

Jung, C.G., *Collected Works*: vol. 7, *Two Essays in Analytical Psychology*, 1953; vol. 8, *The Structure and Dynamics of the Psyche*, 1960; vol. 9, *The Archetypes and the Collective Unconscious*, 1959a; vol. 10, *Civilization in Transition*, 1964; vol. 11, *Psychology and Religion: West and East*, 1958; Routledge & Kegan Paul.

Jung, C.G., *Flying Saucers*, London, Routledge & Kegan Paul, 1959b.

Jung, C.G., *Man and his Symbols*, Aldus Books, 1964.

Juppenlatz, Z., *Urban Wasteland: A Community Resource*, Friends of the Earth, 1981.

Kaplan, R., 'Some psychological benefits of gardening', *Environment and Behaviour*, 5, 2, 1973.

Kaplan, R., 'The role of nature in the urban context', in Altman, I. and Wholwill, J.F. (eds), *Behaviour and the Natural Environment*, Plenum, 1983.

Keates, J.S., *Understanding Maps*, Longman, 1982.

Keeble, D.E., 'Industrial decline, regional policy and the urban-rural manufacturing shift in the United Kingdom', *Environment and Planning A*, 12, 945–62, 1980.

Keeble, D. *et al.*, 'The urban-rural manufacturing shift in the European Community', *Urban Studies*, 20, 3, 1983.

Kelcey, J.G., 'Industrial development and wildlife conservation', *Environmental Conservation*, 2, 1975.

King & Co., 'Industrial Floorspace Survey, 1975–85', 1985.

Kormondy, E.J., *Readings in Ecology*, Prentice-Hall, 1965.

Kramer, S.N., *History Begins at Sumer*, London, Thames & Hudson, 1958.

Kuhn, T.S., *The Structure of Scientific Revolutions*, University of Chicago Press, 1970.

Landsberg, H.E., 'City climate', in *World Survey of Climatology*, vol. 3, Elsevier 1981.

Landscape Research Group, 'The Aesthetics of Landscape', Symposium, 1979.

Laslett, P., *The World We have Lost Further Explored*, Methuen, 1983.

Laszlo, E., *Introduction to Systems Philosophy: Towards a New Paradigm of Contemporary Thought*, Gordon Breach, 1972.

Laurie, I.C. (ed.), *Nature in Cities: Symposium Proceedings*, Landscape Research Group/ Landscape Institute, 1975.

Laurie, I.C. (ed.), *Nature in Cities*, Wiley, 1979.

Lavender, S., *New Land for Old: The Environmental Renaissance of the Lower Swansea Valley*, Adam Hilger, 1981.

Leach, G., *Energy and Food Production*, International Institute for Environment and Development, 1976.

LEEN (London Energy and Employment Network), *The Recycler's Guide to Greater London*, 1985.

Leopold, A., *A Sand County Almanac, With Essays on Conservation from Round River*, Ballantine Books, 1970.

Lessing, Doris, *The Four-Gated City*, MacGibbon & Kee, 1969.

Lévi-Strauss, Claude, *The Savage Mind*, Weidenfeld & Nicolson, 1966.

Levy, G., *The Gate of Horn*, Faber, 1948.

Lewis, C.A., 'Plants and people in the inner-city', *Planning*, 45, 3, 1979a.

Lewis, C.A., 'The harvest is more than vegetables or flowers', in *Community Gardening: A Handbook*, vol. 35, no. 1, of Brooklyn Botanic Garden Record, New York, Spring 1979b.

Lewis, C.A., 'Healing in the urban environment: a person/plant viewpoint', *Journal of the American Planning Association*, July 1979c.

Lewis, C.A., 'Human dimensions of horticulture', *Proceedings of the International Symposium on Urban Horticulture*, New York Botanic Garden, 1983.

Lewis, J.W. (ed.), *The City in Communist China*, Stanford University Press, 1976.

Lindeman, R., 'The trophic-dynamic aspect of ecology', in Hazen, W.E. (ed.), *Readings in Population and Community Ecology*, W.B. Saunders, 1964.

Lipton, M., *Why Poor People Stay Poor: Urban Bias in World Development*, Temple Smith, 1977.

Lobbenberg, S., *Using Urban Wasteland*, NCVO/TCPA, 1981.

London and Cambridge Economic Service, *The British Economy: Key Statistics 1900–1964*, Times Publications, 1964.

Long, J.F., *Population Decentralization in the United States*, US Bureau of Census, 1981.

Lord, J., *Capital and Steam-Power, 1750–1800*, P.S. King & Son, 1923.

Lowe, P. and Goyder, J., *Environmental Groups in Politics*, Allen & Unwin, 1983.

Lowenthal, D. and Prince., H., 'English landscape tastes', *Geographical Review*, 55, 1965.

Lynch, K., *The Image of the City*, MIT Press, 1960.

Lynch, K., *Managing the Sense of a Region*, MIT Press, 1976.

Mabey, R., *Unofficial Countryside*, Collins, 1973.

McAuslan, P., *Urban Land and Shelter for the Poor*, Earthscan, 1985.

McCullough, J., *Meanwhile Gardens*, Calouste Gulbenkian Foundation, 1979.

McHarg, I., *Design With Nature*, Doubleday, 1969.

McLaughlin, C. and Davidson, G., *Builders of the Dawn: Community Lifestyles in a Changing World*, Stillpoint Press, 1985.

McLuhan, M., *Understanding Media*, Routledge & Kegan Paul, 1964.

Maiken, P.T., 'John Jeavons: The mini-farm alternative', *Chicago Tribune Magazine*, 17 May 1981.

Marcus, C.C. and Sarkissian, W., *Housing As If People Mattered*, London, Architectural Press, 1983.

Maringer, J., *The Gods of Prehistoric Man*, Weidenfeld & Nicolson, 1960.

Martin, R.L., 'Britain's slump: the regional anatomy of job loss', *Area*, 14, 4, 1982.

Martin, R.L. and Hodge, J.S.C., 'The reconstruction of British regional policy', *Environment and Planning C: Government and Policy*, 1, 1983.

Martin, V. (ed.), *Wilderness*, Findhorn Press, 1982.

Martin, V. and Inglis, M. (eds), *Wilderness: The Way Ahead*, Findhorn Press, 1984.

Massey, D. and Meegan, R., 'The new geography of jobs', *New Society*, 17 March 1983.

Maund, R., 'The Greater Manchester adventure: an exercise in strategic environmental improvement', *Environmental Education and Information*, 2, 2, 1982.

Mellanby, K., *Can Britain Feed Itself?*, Merlin, 1975.

Mitchell, B.R., *Abstract of British Historical Statistics*, Cambridge University Press, 1962.

MKDC (Milton Keynes Development Corporation), 'Linear park and open space implementation strategy', December 1982.

Mollison, W., *Permaculture One, Permaculture Two*, Tagari, 1978.

Morris, D., *Self-Reliant Cities*, Sierra Club, 1982a.

Morris, D., *The New City-States*, Institute for Local Self-Reliance, 1982b.

Morris, D., *Be Your Own Power Company*, Rodale Press, 1983.

Morris, D. and Hess, K., *Neighbourhood Power: The New Localism*, Beacon Press, 1975.

Morrison, P.A. and Wheeler, J.P., 'Rural renaissance in America?', *Population Bulletin*, 31, 3, 1976.

Moss, Graham, *Britain's Wasting Acres*, Architectural Press, 1981.

Mostyn, B., 'Personal benefits and satisfactions derived from participation in wildlife projects', Nature Conservancy Council, 1980.

Mudrak, C.Y., 'Sensory mapping and preference for urban nature', *Landscape Research*, 7, 2, Summer 1982.

Mumford, L., *The City in History*, Secker & Warburg, 1961.

Munton, R.J.C., *London's Green Belt: Containment in Practice*, Allen & Unwin, 1983.

Nairn, I., *Outrage*, Architectural Press, 1955.

Naisbitt, J., *Megatrends*, Macdonald, 1983.

Nash, R., *Wilderness and the American Mind*, 3rd edn., Yale University Press, 1982.

NCC (Nature Conservancy Council), 'Urban nature conservation and youth employment', 1982.

Needham, J., *Science and Civilization in China*, vol. II, Cambridge University Press, 1962.

Newman, O., *Defensible Space*, Architectural Press, 1973.

NFHA (National Federation of Housing Associations), *Inquiry into British Housing* (chairman, the Duke of Edinburgh), NFHA, 1985.

Nicholls, D.C. *et al.*, 'The private sector housing development process in inner city areas', University of Cambridge, Dept of Land Economy, 1980.

Nisbet, R., *Prejudices – A Philosophical Dictionary*, Harvard University Press, 1982.

North, Marianne, *Recollections of a Happy Life* (1893; republ. as *A Vision of Eden*, Webb & Bower, 1980).

Oakey, R.P., *High Technology Industry and Industrial Location*, Gower, 1981.

Oatley, K., *Perceptions and Representations*, Methuen, 1978.

O'Connor, A., *The African City*, Hutchinson, 1984.

Odum, E.P., *Fundamentals of Ecology*, Saunders College, 1971.

Odum, H.T., 'Energetics of world food productions', in *The World Food Problem: A Report of the President's Science Advisory Committee*, Panel on World Food Supply, White House, 1967.

Odum, H.T., *Environment, Power and Society*, Wiley, 1971.

Oke, T.R., *Boundary Layer Climates*, Methuen, 1978.

Olsen, D.J., *The Growth of Victorian London*, Batsford, 1976.

Opie, I. and P., *Children's Games in Street and Playground*, Oxford University Press, 1969.

Orwell, George, *The Road to Wigan Pier*, Victor Gollancz, 1937.

Otto, R., *Das Heilige*, Breslau, 1917, transl. J.W. Harvey as *The Idea of the Holy*, Oxford University Press, 1923.

Owen, D.F., *What Is Ecology?*, Oxford University Press, 1974.

Owen, J., *Garden Life*, London, 1983.

Park, R.E., Burgess, E.W. and McKenzie, R.D., *The City* (1925), University of Chicago Press, 1967.

Pennick, N., *The Ancient Science of Geomancy*, Thames & Hudson, 1979.

Piaget, J. and Inhelder, B., *The Child's Conception of Space*, W.W. Norton, 1967.

Piaget, J. and Inhelder, B., *The Psychology of the Child*, Basic Books, 1969.

Post, Laurens van der, *The Lost World of the Kalahari*, Hogarth Press, 1958.

Press, I. and Smith, M.E., *Urban Place and Process: Readings in the Anthropology of Cities*, Macmillan, 1980.

Prigogine, I. and Stengers, I., *Order Out of Chaos*, Heinemann, 1985.

Pugh, D.S., *Organization Theory*, Penguin, 1971.

Rackham, O., *Trees and Woodlands in the British Landscape*, Dent, 1976.

Rapoport, A., 'Neighbourhood heterogeneity or homogeneity', *Architecture and Behaviour*, 1, 1, 1980.

Rapoport, A., 'Identity and environment: a cross-cultural perspective', in Duncan, J.S., op. cit, 1981.

Rasmussen, S.E., *London: The Unique City*, revised edn, MIT Press, 1982.

Ratcliffe, D.A., 'Ecological effects of mineral exploitation in the United Kingdom, and their significance to nature conservation', *Proceedings of the Royal Society of London*, A, 339, 1974.

Ravetz, A., *Remaking Cities*, Croom Helm, 1980.

Redfern, P., 'Profile of our cities', *Population Trends*, 30, Winter 1982.

Redfield, R., Singer, M.P. *et al.*, 'The cultural role of cities', *Economic Development and Cultural Change*, 3, 1954–55.

Relph, E., *Place and Placelessness*, Pion, 1976.

Rigby, A., *Communes in Britain*, Routledge & Kegan Paul, 1974.

Riley, P., *Economic Growth: The Allotments Campaign Guide*, Friends of the Earth, 1979.

Robertson, J., *Future Work*, Temple Smith/Gower, 1985.

Robinson, A.H. and Petchenik, B.B., *The Nature of Maps*, University of Chicago Press, 1976.

RSA (Royal Society of Arts), 'Urban Wasteland', Conference, 18 June 1980, Report, *RSA Journal*, November 1980.

RSA, 'Urban Wasteland', Seminar, 12 November 1980, Report, *RSA Journal*, February 1981.

RTPI (Royal Town Planning Institute), *Land Values and Planning in the Inner City*, Report, 1978.

Ruff, A.R., *Holland and the Ecological Landscapes*, Deanwater Press, 1979.

Ruff, A.R. and Tregay, R., 'An ecological approach to urban landscape design', Department of Town and Country Planning, Manchester University, 1982.

Russell, B., *A History of Western Philosophy*, Allen & Unwin, new edn, 1961.

Russell, P., *The Awakening Earth*, Routledge & Kegan Paul, 1982.

Ryan, A., *Property and Political Theory*, Blackwell, 1984.

Rykwert, J., *The Idea of a Town*, Faber & Faber, 1976.

Sames, T., 'Wildlife in towns: a teacher's guide', Nature Conservancy Council, 1982.

Schmitt, P., *Back to Nature: Arcadian Myth in Urban America*, Oxford University Press, 1970.

Schultz, B. and Hughes, D. (eds), *Ecological Consciousness*, University Press of America, 1981.

Schwartz, B., *The Changing Face of the Suburbs*, University of Chicago Press, 1976.

Scientific American, Cities: Their Origins, Growth and Human Impact (Readings), W.H. Freeman, 1973.

Scully, V., *The Earth, the Temple and the Gods*, Yale University Press, 1979.

Seldman, N., 'Why the Synthetic Fuels Corporation should invest in recycling', *Environment*, 25, 2, March 1983.

Seldman, N. and Huls, J., 'Community development and recycling', *Resource Recycling*, January/February 1984.

Sennett, R., *The Fall of Public Man*, Cambridge University Press, 1977.

Sennett, R. (ed.), *Classic Essays on the Culture of Cities*, Appleton-Century-Crofts, 1969.

Sheldrake, R., *A New Science of Life: The Hypothesis of Formative Causation*, Blond & Briggs, 1981.

Shoard, M., 'Metropolitan escape routes', *The London Journal*, 5, 1, 1979.

Shoard, M., *The Theft of the Countryside*, London, Temple Smith, 1980.

Shuttleworth, S., 'The evaluation of landscape quality', *Landscape Research*, 5, 1, 1979.

Sjoberg, G., *The Preindustrial City, Past and Present*, Free Press, 1960.

Skeffington, A. (chairman), *People and Planning*, Report of the Committee on Public Participation in Planning, HMSO, 1969.

Soyka, F., *The Ion Effect*, E.P. Dutton, 1977.

Standing Conference of London and South-East Regional Planning, 'The improvement of London's Green Belt', 1976.

Stearn, Jacqui, *Towards Community Uses of Wasteland*, London, National Council for Voluntary Organizations, 1981.

Strange, P., 'Permaculture', *The Ecologist*, 13, 2/3, 1983.

Sukopp, H. and Werner, P., *Nature in Cities*, Council of Europe, 1982.

Teagle, W.G., *The Endless Village*, Nature Conservancy Council, 1978.

Theodorson, G.A. (ed.), *Urban Patterns: Studies in Human Ecology*, Pennsylvania State University Press, 1982.

Thomas, C., *Material Gains: Reclamation, Recycling and Reuse*, London, 1979.

Thomas, D., *London's Green Belt*, Faber & Faber, 1970.

Thomas, D., 'England's golden west', *New Society*, 5 May 1983a.

Thomas, D., 'Trying to rescue the West Midlands', *New Society*, 4 August 1983b.

Thomas, K., *Man and the Natural World: Changing Attitudes in England 1500–1800*, Allen Lane, 1983.

Tillyard, E.M.W., *The Elizabethan World Picture*, Penguin, 1972.

Todd, N.J. and Todd, J., *Bioshelters, Ocean Arks, City Farming: Ecology as the Basis of Design*, Sierra Club, 1984.

TOES (The Other Economic Summit), *New Economics 85*, Report and Summary, 1985.

Toffler, A., *The Third Wave*, Collins, 1980.

Tolkien, J.R.R., *The Lord of the Rings*, Allen & Unwin, 1959.

Tomkins, P. and Bird, C., *The Secret Life of Plants*, Allen Lane, 1974.

Tooley, R.V., *Maps and Map Makers*, Batsford, 1978.

Towers, B., *Teilhard de Chardin*, Lutterworth Press, 1966.

Toynbee, A.J., *Cities on the Move*, Oxford University Press, 1970.

Tregay, R., 'Urban woodlands', in Laurie, op. cit., 1979.

Tregay, R., *Holland 1980: More (And Even Better) Ecological Landscapes*, Warrington-Runcorn Development Corporation.

Tregay, R. and Gustavsson, R., *Oakwood's New Landscape – Designing for Nature in the Residential Environment*, Warrington-Runcorn Development Corporation, 1983.

Tregay, R. and Moffatt, D., 'An ecological approach to landscape design and management in Oakwood, Warrington', *Landscape Design*, 132, 1980.

Trent, C., *Greater London*, Phoenix House, 1965.

Tuan, Y.F., *Topophilia: A Study of Environmental Perception, Attitudes and Values*, Prentice-Hall, 1974.

US Bureau of the Census, *Indicators of Housing and Neighbourhood*

Quality, Government Printing Office, 1978.

Vine, A. and Bateman, D., 'Organic farming systems in England and Wales: practice, performance and implications', University College of Wales, 1981.

Vining, D.R. and Kontuly, T., 'Population dispersal from major metropolitan regions: an international comparison', *International Regional Science Review*, 1978, 3, pp. 49–73.

Vogtmann, H., 'The quality of agricultural produce originating from different systems of cultivation', transl. Coward, D., Soil Association, 1981.

Walker, D., *The Architecture and Planning of Milton Keynes*, Architectural Press, 1982.

Walker, S.E. and Duffield, B.S., 'Urban parks and open spaces – an overview', *Landscape Research*, 8, 2, pp. 2–12, Summer, 1983.

Walpole, Horace, *Horace Walpole: Gardenist*, ed. Chase I., Princeton, New Jersey, 1943.

Ward, C., *The Child in the City*, Architectural Press, 1978.

Ward, C. and Fyson, A., *Streetwork – The Exploding School*, Routledge & Kegan Paul, 1973.

Wardle, C., *Changing Food Habits in the UK*, Earth Resources Research, 1977.

Wardle, C., *City Farming and Community Gardening 1*, London, Inter-Action, 1983.

Wasteland Forum, *Wasteland Pack*, National Council for Voluntary Organizations, 1982.

Watts, A., *The Way of Zen*, Penguin, 1970.

Weatheritt, D. and Lovett, A., *Manufacturing Industry in Greater London*, Greater London Council, 1975.

Weber, M., *The Religion of China*, The Free Press, 1951.

Weinreb, B. and Hibbert, C., *The London Encyclopaedia*, Macmillan, 1983.

Wells, H.G. and G.P. and Huxley, J.H., *The Science of Life*, 1934.

Wells, T.C.E., 'Creating attractive grasslands', Nature Conservancy Council, 1981.

Wells, T.C.E., 'The creation of species-rich grasslands', in Warren, A. and Goldsmith, F.B. (eds), *Conservation in Perspective*, Wiley, 1983.

Whitehead, A.N., *Process and Reality*, Cambridge University Press, 1927.

Williams, R., *The Country and the City*, Granada (Paladin), 1975.

Williams, R., *Keywords*, Fontana, 1983.

Williams-Ellis, C. (ed.), *Britain and the Beast*, J.M. Dent, 1937.

Wilson, C., *Mysteries*, Hodder & Stoughton, 1978.

Wirth, L., *Urbanism as a Way of Life*, in Sennett, 1969, op. cit.

World Bank, *World Development Report 1984*, Oxford University Press, 1984.

Wray, I., 'Planning for declines? An unconventional regional policy for the 1980s and 1990s', *Regional Studies*, 17, 6, 1983.

Young, K. and Garside, P.L., *Metropolitan London: Politics and Urban Change, 1837–1981*, Edward Arnold, 1982.

INDEX